Design of Internet of Things

T0141281

This book provides a comprehensive overview of the design aspects of the Internet of Things devices and covers the fundamentals of Big Data and Embedded Programming. Various changing technologies, like sensors, RFID, etc., that are promoting IoT solutions are discussed. The core technologies like GSM Communication, Machine to Machine (M2M) Interfaces that are supposed to be the foundations for IoT are explained to prepare the designer. Definitions of IoT, different architectures of IoT, possible models of IOT, along with different market needs and Industry requirements are detailed, so that designers can finalize their choices that enable them to prepare PERT like basics. IEEE Standards useful to IoT design and a few Network Protocol Stacks are explored in detail that will give confidence to any designer. Embedded Operating Systems (FreeRTOS and Contiki) and various Middleware frameworks useful to IoT are discussed. While detailing IoT functional aspects layer wise, the required Memory structures like Ring Buffer Mechanisms are explained, that support designers with IoT Memory management, optimization of software (or code) aspects. Basic code structures and functions required for each layer (e.g. Adaptation Layer, IP, TCP, etc.) are detailed along with required flow diagrams. This will serve as an ideal design book for professionals and senior undergraduate and graduate students in the fields including electrical engineering, electronics and communication engineering, and computer engineering.

This book covers:

- Big Data Aspects that are fundamental drivers to IoT
- Embedded Programming Techniques that are heart for IoT
- Changing Technologies that are promoting IoT solutions
- Machine to Machine Interfaces, the first avatar of IoT
- GSM Communication Aspects and Standards for IoT
- Architecture and different models of IoT
- Conceptual implementation of IoT
- IEEE standards useful to IoT
- Required Embedded Operating Systems suitable for IoT

- Various Middleware Frameworks, for IoT
- Network Protocol Stacks for IoT
- Memory Management aspects of IoT
- Basic functional calls required for each layer of IoT
- Protocol layer wise design concepts useful to IoT (flow diagrams)
- Security concepts for IoT

This book focuses on practical design aspects such as how to finalize a processor (CPU), integrated circuit, and other software and hardware circuit kits available in the market, which operating system to use, etc., in a single volume. This book will be ideal for professionals and graduates from diverse engineering domains including electrical engineering, electronics and communication engineering, and computer engineering, who are perusing IoT Technology as their profession.

Design of Internet of Things

Gunneswara Rao VSSS Kalaga

CRC Press
Taylor & Francis Group
Boca Raton London New York

First edition published 2023
by CRC Press
6000 Broken Sound Parkway NW, Suite 300, Boca Raton, FL 33487-2742

and by CRC Press
4 Park Square, Milton Park, Abingdon, Oxon, OX14 4RN

CRC Press is an imprint of Taylor & Francis Group, LLC

© 2023 Gunneswara Rao VSSS Kalaga

Library of Congress Cataloging-in-Publication Data
Names: Kalaga, Gunneswara Rao VSSS, author.
Title: Design of internet of things / Gunneswara Rao VSSS Kalaga.
Description: First edition. | Boca Raton : CRC Press, 2023. |
Includes bibliographical references and index.
Identifiers: LCCN 2022023491 (print) | LCCN 2022023492 (ebook) |
ISBN 9781032300498 (hbk) | ISBN 9781032300504 (pbk) |
ISBN 9781003303206 (ebk)
Subjects: LCSH: Internet of things.
Classification: LCC TK5105.8857 .K35 2023 (print) |
LCC TK5105.8857 (ebook) | DDC 004.67/8–dc23/eng/20220908
LC record available at https://lccn.loc.gov/2022023491
LC ebook record available at https://lccn.loc.gov/2022023492

ISBN: 9781032300498 (hbk)
ISBN: 9781032300504 (pbk)
ISBN: 9781003303206 (ebk)

DOI: 10.1201/9781003303206

Typeset in Sabon
by Newgen Publishing UK

This work is dedicated to my parents:
Sri Kalaga VSS Sambhu Prasad and Smt Kalaga Sodemma

Contents

15 IoT Security   121

Preface

Internet of Things (IoT) is a growing and much-talked technology world-wide. Government, academia, and industry are involved in different aspects of research, implementation, and business with IoT. IoT domains include healthcare, manufacturing, construction, agriculture, space, water, and mining, to name but a few, which are presently transitioning their legacy infrastructure to support IoT. Today's storage, communication, and computation technologies, in turn, give rise to building different IoT solutions. IoT-based applications such as digital shopping and remote infrastructure management will soon reach both urban and rural areas equally. Remote health monitoring, emergency notifications, and transport applications are gradually relying on IoT-based technology. Therefore, it is important to learn the fundamentals of this emerging technology.

I hope everyone may find it interesting to read "Design of IoT," as an emerging technology. This book is meant for a serious designer, who has understood the basics of computing, communication, programming, etc., and who is interested in attempting to build an IoT device. All the basic fundamental thought processes and requirements related to the design of IoT have been covered, including IoT security. Remember that IoT generates lots of data from the right, left, and center, and hence it will give rise to the growth of "Big Data" (Data Science) technologies. In the introductory part, Big Data concepts and embedded programming techniques are also dealt with (Chapters 1, 2, and 3) for the complete understanding of the intended reader. Chapter 4 deals with the under-laying and changing (improving/growing) technologies like RFID, IC/sensors, M2M, etc. Chapter 5 discusses M2M in detail, to prepare the designer to gain confidence. Chapter 6 deals with the definitions of IoT, while Chapter 7 deals with the introduction to Global System for Mobile Communication and precise details of standards-based IoT architecture models. To make it clear the perceptions of the resultant IoT world, Chapter 8 deals with IoT-based smart city application and also the IoT viewpoints of Indian government.

As a next required step, Chapters 10, 11, and 12 deal with all the basic design requirements like use cases and application scenarios. IEEE standards-based message frames, hardware and software components, and available possibilities that are required for the overall planning of the IoT design approach are covered. Chapter 13 handles all the details of the design requirements and operating systems so that the serious designer can finalize his thought process toward the design of an IoT device. Chapter 14 deals with all the necessary concepts, preparation steps, design models, embedded operating system examples, required code structures, constructs, and international standards that are required for the actual implementation and design of an IoT device. Details of the LBR (6LoWPAN Border Router) which shall be the link between the IoT and its work application server and design aspects are also included. Step by step each designer can implement a targeted IoT device, using and choosing his choices. Chapter 15 talks about IoT security aspects and the future based on IEEE message frames and security touchpoints.

The IoT design community may use the given details and can form a team to scientifically deliver the planned targets in a stipulated time frame. I hope to reach out to every intended designer and equip them in the right direction. I request readers to pardon me for any discomfort.

Acknowledgments

I am grateful to my wife KVS Rama Lakshmi, who supported me very patiently while I worked on this book during the last few months.

I am thankful to Prof. A. Subrahmanyam, IIT Chennai, who read this book and helped me to complete my objectives.

I am also thankful to my daughter Sri Haritha Mantrala and son-in-law Hema Chandra Mantrala, my son Sambhu Prasad Kalaga and my daughter-in-law Sai Lakshmi Lochanam Kalaga, who from time to time helped me with proofreading.

I am thankful to my brothers Dr. KVGS Murty, MBBS, Prof. KVV Atchaiah Sastry, and all my classmates and colleagues because of who I am, what I am now.

Special thanks to the printers and publishers (CRC Press, Taylor & Francis Group) who provided invaluable suggestions, and showed me this book in print.

About the Author

**Gunneswara Rao VSSS Kalaga** is a Dynamic Senior Management professional in computers, communications, and telecom, offering a distinguished and insightful exposure of over 34 years, heading technology administration, project management, product development, and strategy planning with Return of Investment (ROI) accountability. He has spearheaded and functioned as Head of Technical Support (Chief Technology Officer (CTO) Function) with Reliance Communications Limited, Mumbai. He previously worked as the director of D-Link India Limited, and manager of R&D for Electronics Corporation of India Limited (ECIL) at Computer Group. Recognized for his keen analysis and team approach to implement best practices, he is adept at working in high-pressure environments with strict deadlines and multiple deliverables. He is a team manager having the ability to lead cross-functional project teams and integrate their efforts to maximize operational efficiency. He has established credibility in spearheading the entire project management initiatives right from the conceptualization, feasibility studies, preparing financial models, strategic planning, and operational analysis. He has superior communication and interpersonal skills and is multilingual, with proficiency in English, Hindi, and Telugu. He has track records for the attainment of business goals through technology leadership, simple business equations, and management of end-to-end product life cycle of products (hardware and software). He has authored many technical papers published in national and international journals. He is an IEEE member (Membership No. 94277005) and is recognized as a published author.

Chapter 1

Introduction

Much of the learning is from cause and effect. Learning is recognizing or understanding the use of otherwise different scattered thoughts into a concrete useful pattern. This is quite similar to our brain which always depends on the sensory inputs from the eyes, ears, skin (touch and feel), etc., of our body. Children learn a lot from the nature and environment around them like, for example, to touch a physical object or not. And not to touch a hot stove, through that of his friend's experience or through that of his own painful experience; hence, the brain remembers this pattern (it becomes concrete) and that child or person shall be careful to handle any hot object or fire in the course of his life. Ancient society or even recent Vishwa Kavi Rabindranath Tagore and his "Santiniketan" believed in learning through physical interactions with our nature around us or the environment. So-called modern people may be and are more comfortable calling it a Lab Work (Physics Lab, Chemistry Lab, Physical Education, Craftwork, and so on so forth).

Once I asked my child after his third class about what he learned that day, and he said

> We did not learn anything, just we were asked to play with the whole big number of balls with different colors; I did separate one hundred balls each one with different colors; But Ravi could separate only 35, while Rekha did 75, and then you know, the teacher gave me star.

In actual terms, these kids learned a lot of numbers, count, recognition, and so on so forth, all within a day. We may call it "manipulative learning from tangible objects" which is very good. Fortunately, this method is good for learning at any age, and we forget it after some age, thinking that we have learned (of course we learned or not is an individual's paradigm).

Let us talk about the color of the sky—most of us know that the color of the sky is BLUE. In physical terms, there is no specific color in the so-called sky, and every one of us shall receive the reflection of the light from the sky as blue, hence we call it "SKY is BLUE." This is a very useful concrete

DOI: 10.1201/9781003303206-1

phenomenon as every human being on this earth will recognize the sky with blue color only. Hence, we all see in the paints, pictures, photos, movies, and whenever we wish to represent the sky, we always use the color BLUE thoughts ended into a concrete useful pattern.

Another example is humidity at a particular location—the water vapor content in the air at a particular spot/room /building/area—shall be useful in many ways to understand the upcoming atmospheric changes or to design useful air-conditioning equipment, etc.

> Absolute humidity is the total mass of water vapor present in a given volume of air. It does not consider temperature. Absolute humidity in the atmosphere ranges from near zero to roughly 30 grams per cubic meter when the air is saturated at 30 °C.

There are various devices used to measure and regulate humidity. A device used to measure humidity is called a psychrometer or hygrometer. We are trying to identify the humidity with a unique identification pattern to utilize the millions and billions of humidity data for certain scientific calculations or increase human comfort and efficiency.

In this way, if we can represent or recognize each of the objects in this world—a flower, color, location, area, and building—or other requirements like—temperature, water pressure, humidity, etc., with one unique identifiable pattern or process for each thing or object—then one can say that you have become master of the "Internet of Things" (IoT), the emerging technology. All this looks easy, but a bit more complicated and we need to learn more.

We need to remember one thing called IPv6 (Internet Protocol version 6) which provides a 128-bit address mechanism. This means we can uniquely recognize about "340 trillion, trillion, trillion" different objects. As of now, we are about 8 billion (2021 global population) humans, and about 20–50 billion animals (possible Guesstimate from World Atlas 2020), about similar birds and small insects, etc., livestock, 100 billion plant species; maybe about another requirement of 100 billion physical objects including the commercial institutions, etc., together with about less than 500 billion unique present estimate and identifiable maximum objects including humans, livestock, marine stock, etc. (refer to the Table 1.1). Note that we are talking about 500 billion unique objects while we have as many as "340 trillion, trillion, trillion" unique addresses. Hence, if we can systematically work, we may be able to achieve our goal of a "unique identification pattern to all our future and present humans as well as livestock and other objects." We are confident that even we can plan the 256-bit addressing mechanism if need be, but let us not be overenthusiastic, and hence let us not worry about that 256-bit addressing for the present.

Table 1.1 Census of species

Census of Species
Eight million, seven hundred, and four thousand eukaryote species share
this planet, give or take 1.3 million. Eukaryotes have cells with nuclei and
other membrane-bound structures, which means bacteria and other simple
organisms were excluded from the count.

Census Results for the Five Kingdoms of Eukaryotes (approximate)
ANIMALS—7.77 million species (of which 953,434 have been described and
cataloged)
PLANTS—298,000 species (of which 215,644 have been described and
cataloged)
FUNGI—611,000 species (of which 43,271 have been described and
cataloged)
PROTOZOA—36,400 species (single-cell organisms with animal-like
behavior, such as movement, of which 8,118 have been described and
cataloged)
CHROMISTS—27,500 species (including brown algae, diatoms, water
molds, of which 13,033 have been described and cataloged)

Many of the researchers involved in the census look forward to the
discovery of the millions of species yet to be described, but they fear that
many species may disappear before they are even discovered.

Now let us talk about the next complex issue after all the livestock and
human beings, what physical objects and requirements to be identified with?
Most important after livestock in my view are health care and life-system-
related requirements. We should identify all the health and life systems
requirements with unique identification patterns. There are internation-
ally recognized IPv4 and IPv6 Address Allocation Agencies that are already
working on such address allocations, and let us take the support from them.

Chapter 2

Big Data

Great that we could learn a bit of the IoT, let us see the other side of the coin that is called Big Data. For example, about 300 petabytes of data items are being added monthly on Facebook, and every second Picasa album is getting more photos which are even much bigger activities than Facebook. According to the Global Market Intelligence Firm, IDC Corporate, USA, digital data is about eight to ten zettabytes even in 2015, which means we are already dealing with big data volumes. Let us imagine that we shall be getting regular inputs from all the different species, which means about 9 million data items minimum for every second though in principle we shall be getting about 250 Kbps data from every IoT device installed which is a few magnitudes higher. Also, there are humans, corporate, who are already generating the present digital data universally. So, there is a definite connection between IoT evolution and Big Data that we need to understand. Big Data masters should become ready with such mechanisms, so every day's IoT data is instantly analyzed and useful results are separated and stored along with respective data inputs for checking if required in the future, and get ready for a new challenge and new day. Is this not a bit more complicated? Let us see it in due course.

Earlier it was thought that Big Data is only for big corporate like Airlines, Auto, Energy, Financial Services, and Retail. But as per our above observations, the Big Data Spectrum is spreading like ether all over. We should agree that Big Data is touching every aspect of our life, including health and life systems, plant and species control aspects, business inputs for electricity (meters), traffic control/smart city, road policing, electronic surveillance, drone control, and agriculture management, etc., which anyone can say are only a tip of the iceberg.

One of the main aims of Big Data is to deliver focused answers and personalized Management Information System (MIS) useful for the individual as well as the business requirements. This personalized focus gives higher value and causes customers to become more comfortable, hence loyal, and thus more profit to a business. With Big Data every business has a huge opportunity to present customers with personalized service and

DOI: 10.1201/9781003303206-2

personalized promotions deals and recommendations; but to deliver such clarity, Big Data needs big inputs of data, and the only such focused and clear action is possible through IoT-enabled devices (with micro and miniature sensors which can reflect continuous and constant data input) and can be installed in millions and billions if need be. Hence, it is natural that IoT and Big Data should go hand in hand and deliver value to us.

The kind of focused answers Big Data needs to deliver depends on the data items which we load for getting the results, but the present-day user activity runs into terabytes of data, justifying that more data can deliver correct and focused answers, that is, useful patterns. The traditional databases will be struggling to handle such data and there shall be magnitudes of difference and latency to deliver the right results. Big Data systems need a whole new kind of data handling capability.

We should not forget that the "datum" measure in the digital language is bit, byte, kilobyte, megabyte, gigabyte, terabyte, petabyte, exabyte, zettabyte, and yottabyte. It may take about 1.3 seconds to read 500 kilobytes on a standard high-speed disk (50 Mille Seconds Read Head Access Latency; 5 Mille Seconds Sector positioning [63 Sectors Ave.] and about 500 kilobytes take about 1.3 seconds). Hence, if we wish to read one petabyte of data at a similar speed, it shall take about a minimum of 30 days or so to say 1 month.

Data Science is the concept developed for handling Big Data, and, for example, "Hadoop" and Apache's "Spark" are possible frameworks to handle such a big volume of ever-increasing data. Big Data Analytics is all about the management and analysis of data to a specific requirement. Frameworks like Hadoop, Spark, Flink, Storm, etc., can handle the Big Data that is evolving and current science.

Big Data engineers talk in general about Hadoop and Spark, but the holistic approach towards Big Data is to use the inner layers of the mechanisms like MapReduce, Pig, Hive, YARN, etc., judiciously as they are continuously evolving. In the present-day environment, engineers can use Hadoop tools with Spark Processing Engine and also take advantage of HDFS (Hadoop Distributed File System). Even Cloudera, the well-known Big Data company, is planning the replacement of MapReduce with Spark as an example of how the Big Data tools are evolving continuously and toward focused utility. The other framework "Flink" is based on data flow architecture and has some advantages on row handling while Strom focuses on real-time distributed computing aspects.

Big Data presents new opportunities for some long-standing enterprise solutions. As HP's Srinivasan Rajan points out, COBOL offers several advantages when it comes to handling large data sets. For one, Big Data analytics can take advantage of COBOL's batch infrastructure support and complex algorithms. Also, COBOL's Job Control Language (JCL) is highly adept in scheduling large jobs into smaller ones, helping to lessen the impact on underlying resources. Table 2.1 shows expected data loss based on the

Table 2.1 Acceptable data loss in bytes for 1 PB and 50 PB

One petabyte (PB) in terms of bytes	9s	Data reliability %	1 PB	50 PB
1 125 899 906 842 624 bytes	2	99%	9,00,71,99,25,47,410	4,50,35,99,62,73,70,500
1 099 511 627 776 kilobytes (KB)	3	99.90%	90,07,19,92,54,741	45,03,59,96,27,37,050
1 073 741 824 megabytes (MB)	4	99.99%	9,00,71,99,25,474	4,50,35,99,62,73,700
1 048 576 gigabytes (GB)	5	99.999%	90,07,19,92,547	45,03,59,96,27,350
1024 terabytes (TB)	6	99.9999%	9,00,71,99,255	4,50,35,99,62,750
	7	99.99999%	90,07,19,925	45,03,59,96,250
	8	99.999999%	9,00,71,993	4,50,35,99,650
Prefixes as powers of 1024				
$1024^1 = 1024$ (Kilo)	9	99.9999999%	90,07,199	45,03,59,950
$1024^2 = 1\ 048\ 576$ (Mega)	10	99.99999999%	9,00,720	4,50,36,000
$1024^3 = 1\ 073\ 741\ 824$ (Giga)	15	99.9999999999999%	9	450
$1024^4 = 1\ 099\ 511\ 627\ 776$ (Tera)				
$1024^5 = 1\ 125\ 899\ 906\ 842\ 624$ (Peta)				
$1024^6 = 1\ 152\ 921\ 504\ 606\ 846\ 976$ (Exa)				

number of 9s reliability and data loss in bytes. For example, for ten 9s reliability and one Petabyte of data, the acceptable loss is 900,720 bytes.

Let us see the Data Science requirements example—there are about 23 million telephone calls happening across India every day (Max Call Busy Second is 1200). If any government wishes to learn about a particular issue—per se identified about 32 suspect numbers—one needs to identify the call pattern received for each of the numbers. But the incoming call may come from anywhere, in India or outside India. Data: Every day there are 23 million calls; therefore, monthly data is about 23 million × 30 = 690 million calls—nice Big Data.

Possible patterns, planning, and logical approach for the problem at hand:

- 32 numbers to be compared and pattern match to be done with 690 million Call Data Records—each number one time; or separate 32 number of Server clusters may be assigned to the task so that they happen parallel.
- The estimate is that all the calls from different numbers may originate to one number at a specific time or within 30 minutes or one hour etc., a most generic and possible analogy.
- Each day's data can be divided into 24 parts (based on time and we call it hourly data), and 23 million calls per day divided by 24 hours = now the hourly data is about less than 23,000,000/24 = ~ 959 K data (please note that the actual data division shall be done on time stamp base, and real data shall be at least 8 to 16 times bigger than this [note that we have taken only one data item to represent a call, for easy understanding]). Remember we could reduce it to about 959 K per one Server cluster and we can submit it to such 32 Server clusters for the pattern match with each of the 32 suspect numbers one by one, for a match.

Hope we could get a fair amount of understanding of basics about and Data Science or the so-called Big Data is doable provided that we understand it. There are many paradigms and languages like "R", Python," etc., and many other platforms, to handle Big Data, which are evolving continuously. For the sake of completion, we introduced the detailed concept of Big Data here, and further in this book, we shall not deal with Big Data as we wish to enhance our understanding of IoT and then design the IoT devices (Table 2.2).

Table 2.2 Programming Big Data—concepts

Programming Big Data for Targeted Results:

Per se, Each Server Cluster is of 8 nodes, and hence 959 K data is subdivided into 8 parts to be submitted to each server (959000 / 8 = 119.8 K data Records for each node). There are 32 server clusters of each having 8 nodes i.e. a total of 32 x 8 = 256 nodes shall be working parallel on this single problem; One can plan them as 256 threads; if each Machine has in-built 8 processors (CPU) to handle the issue, with a 32 such Machines in closely coupled network—or parallel computing environment—etc. can be discussed. So, remember we got the data into a manageable format by the way of applying our required rules and pattern matching techniques. Now because we are talking 120 K call data on which we need to apply for the logic and program. The whole problem is can be handled easily, provided one has the desired environment.

Now the programming part / the "ART" — "Mat lab Version 6" has required algorithms (more than one method can be applied) related to Collaborative Filtering Techniques—which can be applied to pattern matching and to suit our above problem requirements. At this moment we shall not get into details of what specific equations or algorithms are used, as at this moment conceptual understanding is of utmost importance.

Result: For initial working it may take about one or two months, to finalize the Algorithms; program error correction; trial and error; debugging; etc. — a learning curve. But once all these are finalized, it shall become routine submission of job to the 32 Server Parallel Computing Environments, and within a flat 7 days; one can send the results to the Government about those specific 32 numbers.

Note: A trained and seasoned team may be able to deliver the same results within 48 hours, provided the environment and the problem definitions are routine to them. That is, they are used to address such problems. We may also call them a "Designed Thinking" team. Data Science or Big Data is a concept or Art so that one needs to understand the problem definition and the result requirements—more often called as, Management of Information—in a simple way.

Chapter 3

Embedded Programming Techniques

Even though it is assumed that every designer is familiar with embedded programming, we thought as part of the introduction we need to present some of the embedded concepts at this juncture. Generally, every Embedded System is unique and the hardware and software shall become highly specific to the identified Application Domain. Today, almost every household has one or other type of Embedded System in use (smart phone, router, ADSL CPE, modem, video camera, Video Cassette Recorder/Reader [VCR], etc.). If an Embedded System is designed well, then the existence of a processor and related software would not be noticed by the user of the device (refer Table 3.1).

At present, there are many proven and goal-oriented embedded development tools and frameworks. To get an idea, a few tools are presented here:

- C/C++/C++ Compilers: Wind River Compiler, Green Hills Compilers, GNU Compilers, etc.
- Real-Time Operating System: Nucleus, VxWorks, Embedded Linux, FreeRTOS, Contiki, etc.
- Make Utilities: OPUS MAKE, GNU MAKE, Microsoft MAKE.
- Debuggers: Monitors, ICE, Simulators, Logic Analyzer, Scope.
- Drivers: All the necessary, Peripheral Device Drivers(S/W).
- Basic Format Conversion Software: (HEX to Binary etc.)
- Hardware: PROM Programmers, IC Testers, CPU Test Kits.

Presently each manufacturer is providing the platform environment and required tools even for the IoT designers.

Simplified "target embedded product development" steps:

Step 1: One has to understand the functional requirements and accordingly should choose the hardware elements like CPU, Peripherals, Bus Controllers, Network Controllers, etc. Now the required device drivers, CPU start-up programs, protocol software, etc., may have to be compiled separately. Based on the target functional data flow

DOI: 10.1201/9781003303206-3

Table 3.1 Embedded system basic requirements

Embedded Systems Technology: Basic concept review
Processor Power: 25MIPS to 40 MIPS, (16 bits, 32 bits, 64 bits). EX: ARM7, ARM11, etc. **Memory: ROM, RAM, FLASH (1MB to 16 MB, 32MB, 64 MB).** **Expected Life Time: one Year to Five Years** **Development Cost: 10,000 to 100,000 US$**

one has to generate/create the source code for each of the identified hardware elements and compile the respective Object Code files for the same. At present, there are many frameworks (Integrated Embedded Development Environments) for such generation of object code files.

Step 2: One has to link all these available object code files related to different hardware elements and software protocols that control the behavior of the final target system into a single, relocatable, final target "Object Code," using appropriate "Linker." Integrated Embedded Development Environments may support the generation of object code files as well the linking of the separate object files into a single relocatable code. One has to take care of the issue as these Integrated Frameworks sometimes may generate additional unnecessary memory gaps or not useful code, which may not affect the target functionality but increase the code size.

Step 3: Note that a single relocatable object code shall appropriately be placed into the final target FLASH/RAM/ROM parts of the target system which shall give the required functional behavior for the target embedded product, on activation. We need to use the "Locator" to place the final code into the respective identified Fixed Memory addresses (target FLASH/RAM/ROM parts of target system) in the form of fixed "binary code." One may call it (final binary code) as final targeted Text Section, Data Section, Stack section, so on and so forth or Final Fixed Target Code. This code can be easily moved into the appropriate FLASH/RAM/ROM parts of the target embedded platform. Now let us get into the programming or the actions part with the available components. Target CPU-based start-up code, which may also be known as Power on Self-Test Code (POST), should contain the following (Table 3.2):

Table 3.2 Embedded system POST steps

* **Disable all the CPU interrupts**
* **Copy initial data from ROM to RAM**
* **ZERO the uninitialized data area in the RAM**
* **Allocate space and initialize the stack in RAM**
* **Initialize the CPU Stack Pointer (put the right address)**
* **Create and initialize Heap area**
* **Execute the constructs for Global variables**
* **Enable the interrupts**
* **Call the "main" program**

The Memory Map and Input and Output (I/O) Map have to be drawn for the target. This shall be the lifeline of Embedded Programmer. The Control Register Addresses, Interrupt Levels, In-built Memory blocks of each part (if any), should be very thorough. Based on the above information one can start writing small diagnostic programs for the target hardware to check each of the identified chips/MCU of this platform is functioning correctly and ready to take runtime software to get the target behavior.

Avoid the use of complete code related to Standard Library Routines, but use only the required part of the code out of their source, whenever possible. Always recognize and remember the native word size of the MCU should be used for "integer declaration" frequently to save many more bytes of code generation.

C/C++ keyword "volatile" should be used to declare any of the Device Registers. Then at the optimization phase, the compiler treats that variable as though its behavior cannot be predicted at compile time. Embedded programmers can divide each part of the code into different possible tasks and manage them through RTOS. (task states—Running, Ready, Waiting, etc.). Take care of the "critical section" of each "task". Interrupts, if any, are to be disabled and enabled properly.

C/C++ keyword "register" may be used for declaring local variables that enable the compiler to place them in a general-purpose register rather than in a stack. Use "Global variables" to pass a parameter to a function. But Global variables may have some negative effects. Yet at times one has to choose between Interrupt and Polling. Sometimes polling may prove better.

Note: Never make the mistake of assuming that optimized code shall behave the same way as the earlier unoptimized code. One has to test the new code again. Figure 3.1 shows the software architecture of the embedded target system.

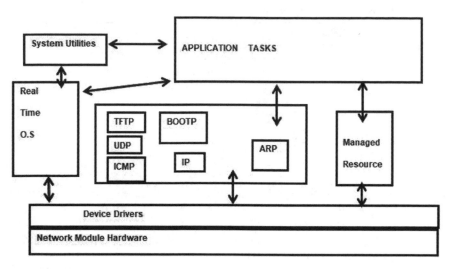

Figure 3.1 Embedded system architecture.

Chapter 4

Changing Technologies

The technology trend is shifting toward providing faster data rates and lower latency connectivity (see the Figure 4.1), the Third Generation Partnership Project (3GPP) "standards body" has developed a series of enhancements to create the "High-Speed Packet Access [HSPA] Evolution," also referred to as "HSPA+." The HSPA Evolution represents a logical development of the Wideband Code Division Multiple Access (WCDMA) approach and is the stepping stone to an entirely new 3GPP radio platform called 3GPP Long-Term Evolution (LTE). LTE offers several distinct advantages such as increased performance attribute(s), high peak data rates, low latency, and greater efficiencies in using the wireless spectrum.

Low latency makes it possible for IoT applications to query or receive quicker updates from sensor devices. LTE networks have latencies on the order of 50–75 msec, which will open up new types of programming possibilities for application developers. For example, wearable computers that require interactive and real-time feedback will require moving large chunks of data to be analyzed in the cloud or back-end systems to create a seamless user experience.

Higher peak data rates can support applications such as Voice over Internet Protocol (VoIP) and digital video that require better quality of service (QoS). With further advancements in communication technologies such as Software Defined Radio (SDR) and "Long-Term Evolution – Advanced" (LTE-A), devices will be able to communicate with better QoS and support better access to new services with more efficient use of the radio frequency spectrum.

4.1 RFID

Radio Frequency Identification (RFID) technology is of particular importance to IoT as one of the first industrial realizations of IoT is in the use of RFID technology to track and monitor goods (i.e., things) in the logistics and supply chain sector. RFID frequency bands range from 125 kHz (low

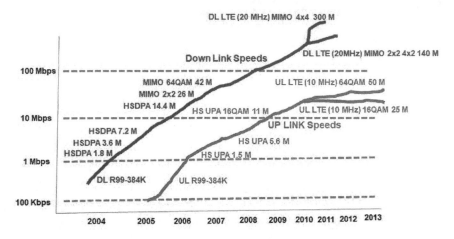

Figure 4.1 UP link and DOWN link speeds.

frequency [LF]) up to 5.8 GHz (super high frequency [SHF]) and the tags have at least three basic components:

- The chip holds information about the object to which it is attached and transfers the data to the reader wirelessly via an air interface.
- The antenna allows transmission of the information to/from a reader.
- The packaging encases the chip and antenna and allows the attaching of the tag to an object for identification.

4.2 IC/Sensor Cost

Just as the size of the chips is getting smaller, the costs of sensing components are also dropping to become more affordable. Gartner has forecast that most technology components such as radio, Wi-Fi, sensors, and global positioning systems (GPS), could see a drop in cost of 15%–45% from 2010 to 2015, and so on so forth. To illustrate, with cheaper temperature sensors, cold chain retailers would consider deploying more temperature sensors to monitor their perishable produce as it traverses the supply chain.

4.3 6LoWPAN

6LoWPAN is an acronym for "IPv6 over Low power Wireless Personal Area Networks." It is a communication standard that allows low-power devices to communicate and exchange data via IPv6. There are many benefits of using IP-based connectivity to form the sensor access network:

- IP connects easily to other IP networks without the need for translation gateways or proxies.
- IP networks allow the use of existing network infrastructure.
- IP is proven to work and scale. Socket API is well-known and widely used.
- IP is open and free, with standards, processes, and documents available to anyone.
- IP encourages innovation and is well understood.

6LoWPAN works on the IPv6 protocol suite based on IEEE 802.15.4 standard. Hence, it has the characteristics of low-cost, low-rate, and low-power deployment. The bottom layer adopts the physical (PHY) and media access control (MAC) layer standards of IEEE 802.15.4 and uses IPv6 as the networking technology.

It is yet again proven that now and then a new technology like IoT will emerge that has the potential to disrupt markets and create many new business opportunities. IoT is poised to disrupt a multitude of industries across the globe, create new business processes, and create an abundance of opportunities for those who can see the technical advances in line with new possibilities. It is to be noted that in this document 6LoWPAN and LoWPAN are both acronyms that are used frequently for the same.

4.4 M2M, IoT, and IoE Concepts

The term "Internet of Things" coined by British entrepreneur Kevin Ashton in 1999 described connectivity among physical objects and no longer holds in its original form. It is now largely mixed-up, confused, and even mystified with the term "Internet of Everything" (IoE). IoE is considered a superset of IoT and machine-to-machine (M2M) communication is considered a subset of IoT. Let's take a closer look into the differences between IoT, IoE, and M2M, which have impacted consumers and businesses alike.

Although the concept of the IoE emerged as a natural development of the IoT movement and is largely associated with Cisco's tactics to initiate a new marketing domain, IoE encompasses the wider concept of connectivity from the perspective of modern connectivity technology use cases.

IoE comprises four key elements including all sorts of connections imaginable:

- People: Considered as end nodes connected across the Internet to share information and activities. Examples include social networks, health and fitness sensors, among others.

- Things: Physical sensors, devices, actuators, and other items generating data or receiving information from other sources. Examples include smart thermostats and gadgets.
- Data: Raw data is analyzed and processed into useful information to enable intelligent decisions and control mechanisms. Examples include temperature logs converted into an average number of high-temperature hours per day to evaluate room cooling requirements.
- Processes: Leveraging connectivity among data, things, and people to add value. Examples include the use of smart fitness devices and social networks to advertise relevant healthcare offerings to prospective customers.

IoE establishes an end-to-end ecosystem of connectivity including technologies, processes, and concepts employed across all connectivity use cases. Any further classifications—such as Internet of Humans, Internet of Digital, Industrial Internet of Things, communication technologies, and the Internet itself—will eventually constitute a subset of IoE if not considered as such already.

The aptly named IoT subset M2M initially represented closed, point-to-point communication between physical-first objects. The explosion of mobile devices and IP-based connectivity mechanisms has enabled data transmission across a system of networks. M2M also more recently is being referred to as technology that enables communication between machines without human intervention. Examples include telemetry, traffic control, robotics, and other applications involving device-to-device communications.

4.5 IoT—Internet of Things

The IoT (i.e., all kinds of M2M devices, virtual devices) holds significant promise for delivering social and economic benefits to emerging and developing economies. This includes areas such as sustainable agriculture, water quality and use, healthcare, industrialization, and environmental management, among others. As such, IoT holds promise as a tool in achieving the United Nations Sustainable Development Goals.

The broad scope of IoT challenges will not be unique to industrialized or developed countries. Developing regions also will need to respond to and address IoT challenges to realize the potential benefits of IoT. Besides, the unique needs and challenges of implementation of IoT in less-developed regions shall equally help to the quick development and commercialization of IoT hence need to be addressed, few pointers in this direction are infrastructure readiness, market and investment incentives, technical skill requirements, and policy resources, etc.

This set of Internet of Things technologies are realizing a vision of a miniaturized, embedded, automated environment of devices communicating constantly and automatically. However, connecting up devices or robots (whether they are bridges, fridges, or widgets) is only a means to an end—the really interesting part arises in terms of what can be done with the data obtained, and the learning outcomes for improving our future.

Houlin Zhao, ITU Secretary-General

Today, the IoT is improving the day-to-day lives of citizens around the world. In cities from Barcelona to Chandigarh to Rio de Janeiro, IP-connected sensors are monitoring traffic patterns, providing city managers with key data on how to improve operations and communicate transportation options. Similar information flows are improving hospitals and healthcare systems, education delivery, and basic government services such as safety, fire, and utilities. Sensors and actuators in manufacturing plants, mining operations, and oil fields are also helping to raise production, lower costs, and increase safety.

Chapter 5

M2M

Through automated money transfer supporting machines like automatic teller machines (ATMs), we could save considerable resources, which are providing more security to critical sectors. Similarly, M2M applications will support us to master the demographic changes. Around the mid-1990s when the first of the modems came to the market, before our eyes, we saw the increase of circuit-switched data transfer standards from speeds of 2400 to 4800 to 9600 baud. Think about the new services and Short Message Service (SMS) standards from General Packet Radio Service (GPRS) to the current LTE/4G/5G. When things become smart in this way, they open up entirely new opportunities.

We can already see this with, say, vending machines. A vending telemetry solution begins by supporting day-to-day business. Operators can check filling levels and operating data remotely and thereby reduce filling and maintenance costs. New services like mobile payment and digital signage are integrated and connections with social media channels are established. But the data collected holds the greatest potential. Operators can see exactly which products are in demand when and where. This knowledge provides them with a new basis on which to make business decisions.

M2M communication is on the rise. There will be more machines connected to the Internet than human beings in the next decade. M2M technologies transfer data on the condition of physical assets and devices to a remote central location for effective monitoring and control. While M2M concepts and technologies have been in use for quite some time, the changing business scenarios and newer use cases are acting as growth stimulants (refer Figure 5.1). Greater demand for M2M solutions is primarily being triggered by the widespread adoption and proliferation of affordable wireless communication.

M2M applications are now part of our daily life, are not something we need to highlight. The concept of the IoT has meanwhile established itself. This in itself proves the new understanding and immense potential when one talks about the networked economy. Regardless of this, when one regards the circulated figures of networked machines and equipment, it is clear that

DOI: 10.1201/9781003303206-5

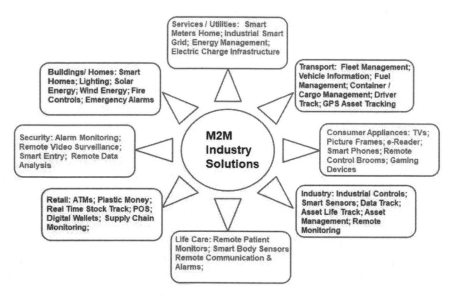

Figure 5.1 M2M industry solutions.

we will be networking even more elements with one another, soon. These intelligent solutions will help us to simplify our lives further.

The European Telecommunications Standards Institute (ETSI) defines M2M communication as follows: "Machine to Machine (M2M) communication is the communication between two or more entities that do not necessarily need any direct human intervention. M2M services intend to automate decision and communication processes."—Remember this is one of the many definitions that are floating around us, from different bodies.

To take advantage of the opportunities M2M presents, entrepreneurs are looking into the big picture to find novel solutions to mass usage business environments. In the current environment, M2M can be used in almost all aspects of life. If we look at government setup across various ministries and departments, more than 70% of them are either using M2M or planning to use M2M technologies in one way or other. With better sensors, wireless networks, and increased computing capability, deploying an M2M makes sense for many sectors. "M2M" has opened up many such opportunities in technology and business, that its myriad applications extend beyond the corporate world into our daily lives and have transformed the way we live, work, and play. Let's have a look at just a few of the possible market opportunities from M2M technology in three different industries:

Construction: With M2M, everything that can be connected will be connected. Future buildings are no exception. In an age of smart grids, buildings are being equipped with thousands of sensors to monitor, control, and optimize everything. There's a tremendous opportunity for entrepreneurs to reinvent traditional products like connected plugs, light switches, and heating/cooling vents that can make predictive "decisions" by anticipating energy needs without human intervention.

Transportation: The transportation industry is also rich with "M2M" opportunities. Several entrepreneurs have already started services like Fleet Management, Radio Taxi, and smartly managed logistics. Further, app developers could create smartphone apps for vehicles with wireless connections to remotely monitor and control fuel consumption, locate petrol pumps or pre-cool a car by remotely switching on AC based on past usage.

Agriculture: There is always a rising demand for food and reducing agricultural land that puts upward pressure on all food input costs. There are novel opportunities for M2M solutions in the agriculture industry, especially for raw materials and energy. Connected devices can help maximize production efficiency and yield to improve the grower's profit margins. For example, entrepreneurs could build auto-pilot tractors that automate planting and plowing solutions to reduce labor costs, fuel, and waste. Furthermore, wireless integration of farm vehicles with farm management software could integrate important factors such as tractor usage and crop yield information.

Mobile network coverage is being expanded worldwide. Telemetries are seen increasingly as sources of greater operational efficiency and increased incremental revenue.

- **M2M applications benefit from R&D and the scale of the mobile handset industry.**
- **Technical advances in air interface standards are enabling new telecom M2M market segments.**
- **Mobile Network Operators (MNOs) are seeking to expand their data service offerings.**
- **Government mandates are increasingly requiring the use of telemetries and related functionality.**

As MNOs become more directly involved with M2M application service providers (ASPs), many are deploying key network elements, specifically mobile packet gateways (e.g., Gateway GPRS Support Node—GGSN; Packet Data Serving Node—PDSN; etc.), specifically for their M2M operations, separate from their general mobile data infrastructure. Key benefits of doing this include simplification of internal business operations and optimization of network utilization.

There was an estimated half a billion devices connected in 2012, and the trend is growing 45%–50% annually. According to different audited report indications, the global mobile M2M market was to reach $57 billion by 2014. All types of M2M connections were tripled—reaching more than 225 million connections 2014.

With the expected increase in the deployment of mobile network elements dedicated specifically to M2M applications, likely, such equipment will increasingly be designed for the specific needs of M2M applications. For example, mobile packet gateways that are optimized for M2M traffic are designed to handle a large number of packet data sessions, rather than provide large amounts of throughput. Fundamentally, there is a range of technical parameters upon which mobile packet gateways can be optimized for M2M application deployment.

5.1 M2M Architecture

"M2M" communication can also be carried over mobile networks (e.g., GSM-GPRS and CDMA EVDO networks). Note that, for M2M communication, the role of the mobile network is largely confined to serving as a transport network only. The areas in which M2M is currently in use are listed briefly here:

- Security: Surveillances, Alarm systems, Access control, Car/driver security
- Tracking & Tracing: Fleet Management, Order Management, Pay-as-You-Drive, Asset Tracking, Navigation, Traffic Information, Road Tolling, Traffic optimization/steering
- Payment: Point of sales, Vending machines, Gaming machines
- Health: Monitoring vital signs, Supporting the aged or handicapped, Web Access Telemedicine points, Remote diagnostics
- Remote Maintenance/Control: Sensors, Lighting, Pumps, Valves, Elevator control, Vending machine control, Vehicle diagnostics
- Metering: Power, Gas, Water, Heating, Grid control, Industrial metering
- Manufacturing: Production chain monitoring and automation
- Facility Management: Home/building/campus automation

M2M applications will be based on the infrastructure that is provided by one or many communication service providers (CSPs). Applications may either target end-users, such as a user of a specific M2M solution, or other application providers to offer more refined building blocks by which they can build more sophisticated M2M solutions and services (refer Figure 5.2). Examples include customer care functionality, elaborate billing functions, etc. Those services, or service enablers, may be designed and offered by an

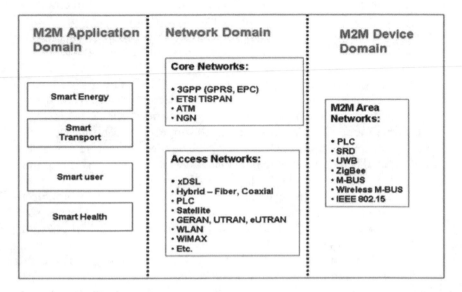

Figure 5.2 M2M architecture.

application provider, but they might be offered by the operator via the operator platform itself.

M2M features include the following:

- Low Mobility: M2M devices are supposed to be very small and they need not move. People may move with them (Roaming facility).
- Time Controlled: Possible to send or receive data only at certain predefined periods.
- Packet Switched: Network operator to provide packet-switched service.
- Small Data Transmissions: M2M devices send or receive small amounts of data.
- Monitoring: Provide functionality to monitor the events.
- Low Power: Use very low power to improve the system and efficiency of service.

5.2 Key Technical Requirements of M2M

M2M Application communication principles: The M2M system shall be able to allow communication between M2M applications in the Network and Applications Domain, and the M2M device or M2M gateway, by using multiple communication means, e.g., SMS, GPRS, and IP Access. Also, a Connected Object may be able to communicate in a peer-to-peer manner with any other Connected Object. The M2M system should abstract the underlying network structure including any network addressing mechanism used; for example, in the case of an IP-based network, the session establishment shall be possible when IP static or dynamic addressing is used.

Message delivery for sleeping devices: The M2M system shall be able to manage communication toward a sleeping device. (One of the methods is—the message is delivered whenever the device is live.)

Delivery modes: The M2M system shall support anycast, unicast, multicast, and broadcast communication modes. Whenever possible a global broadcast should be replaced by multicast or anycast to minimize the load on the communication network.

Message transmission scheduling: The M2M system shall be able to manage the scheduling of network access and of messaging. It shall be aware of the scheduling delay tolerance of the M2M application.

Message communication path selection: The M2M system shall be able to optimize communication paths, based on policies such as network cost, delays, or transmission failures when other communication paths exist.

Communication with devices behind an M2M gateway: The M2M system should be able to communicate with devices behind an M2M gateway.

Communication failure notification: M2M applications, requesting reliable delivery of a message, shall be notified of any failures to deliver the message.

Scalability: The M2M system shall be scalable in terms of many Connected Objects.

Abstraction of technologies heterogeneity: The M2M gateway may be capable of interfacing with various M2M Area Network technologies.

M2M service capabilities discovery and registration: The M2M system shall support mechanisms to allow M2M applications to discover M2M Service Capabilities offered to them. Additionally, the M2M device and M2M gateway shall support mechanisms to allow the registration of its M2M Service Capabilities to the M2M system.

M2M trusted application: The M2M Core may handle service request responses for trusted M2M applications by allowing streamlined authentication procedures for these applications. The M2M system may support trusted applications that are applications pre-validated by the M2M Core.

Mobility: If the underlying network supports seamless mobility and roaming, the M2M system shall be able to use such mechanisms.

Communications integrity: The M2M system shall be able to support mechanisms to assure communication integrity for M2M services.

Device/Gateway integrity check: The M2M system shall support M2M device and M2M gateway integrity checks.

Continuous connectivity: The M2M system shall support continuous connectivity, for M2M applications requesting the same M2M service on a regular and continuous basis. This continuous connectivity may be deactivated upon request of the application or by an internal mechanism in the M2M Core.

Confirm: The M2M system shall support mechanisms to confirm messages. A message may be unconfirmed, confirmed, or transaction controlled.

Priority: The M2M system shall support the management of priority levels of the services and communications services. Ongoing communications may be interrupted to serve a flow with higher priority (i.e., preemption).

Logging: Messaging and transactions requiring nonrepudiation shall be capable of being logged. Important events (e.g., received information from the M2M device or M2M gateway is faulty, unsuccessful installation attempt from the M2M device or M2M gateway, service not operating, etc.) may be logged together with diagnostic information. Logs shall be retrievable upon request.

Anonymity: The M2M system shall be able to support anonymity. If anonymity is requested by an M2M application from the M2M device side and the request is accepted by the network, the network infrastructure will hide the identity and the location of the requestor, subject to regulatory requirements.

Time Stamp: The M2M system shall be able to support accurate secure and trusted time-stamping. M2M devices and M2M gateways may support accurate secure and trusted time-stamping.

Device/Gateway failure robustness: After a nondestructive failure, for example, after a power supply outage, an M2M device or gateway should immediately return in a full operating state autonomously, after performing the appropriate initialization, for example, integrity check if supported.

Radio transmission activity indication and control: The radio transmitting parts (e.g., GSM/GPRS) of the M2M device/gateway should be able to provide (if required by particular applications, e.g., eHealth) a real-time indication of radio transmission activity to the application on the M2M device/gateway, and may be instructed real-time by the application on the M2M device/gateway to suspend/resume the radio transmission activity.

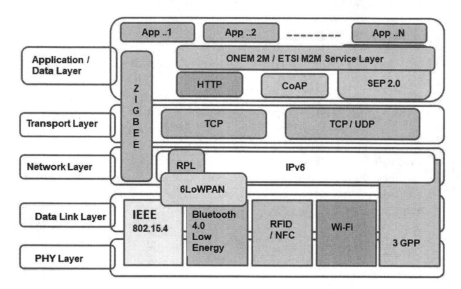

Figure 5.3 M2M protocol stack.

Figure 5.3 presents the M2M protocol stack as viewed by many CSPs and application vendors.

Chapter 6

Definitions of IoT

Some people define the IoT as a shorthand way of describing a globally interconnected continuum of devices and objects interacting with the physical environment, people, and each other. And the subset of IoT is M2M communication that describes a set of interconnected devices that allow both wireless and wire-line systems to communicate with other devices, mostly in vertical segments, that is, the plumbing/connectivity that enables the IoT ecosystem.

The social and individual benefits of the IoT are realized with M2M innovations. More than just providing the convenience of the Internet, the IoT will provide greater efficiency by automating tasks, exchanging information, performing updates, making adjustments, maintaining thresholds, and comparing variances. Machines will communicate directly with one another based on intelligent algorithms that help liberate us from routine tasks, improve end-user quality of life, reduce complexity and cycle time, improve efficiency and often enhance safety.

Four main categories of technology advances are contributing to the rapid rise of IoT and M2M communications (refer Figure 6.1):

Tagging Things—Technologies responsible for Radio Frequency Identification (RFID), Near-Field Communication (NFC), Quick Response (QR) codes, and Digital Watermarking.

Sensing Things—Technologies that react to environmental conditions, such as the presence of moisture (smart textiles, smart pavement, and water leak detectors), heat (smart thermostats), and air quality (smoke alarms, HVAC monitors, etc.).

Shrinking Things—Technologies that make IT objects smaller, lighter, and smarter, where embedded computing and wireless are the new norms.

Thinking Things—Objects that access the semantic Web, which supports standardized formats to enable people to share content beyond the boundaries of applications and websites, and open cloud data to customize things.

DOI: 10.1201/9781003303206-6

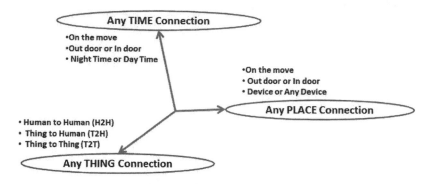

Figure 6.1 IoT architecture.

International Telecommunications Union (ITU) endorses the definition of IoT as a network that is available anywhere, anytime, by anything and anyone. Hence, connectivity will take on an entirely new dimension. Today, users can connect at any time and from any location. Tomorrow's global network will not only consist of humans and electronic devices but also all sorts of inanimate things. These things will be able to communicate with other things, for example, fridges with grocery stores, laundry machines with clothing, implanted tags with medical equipment, and vehicles with stationary and moving objects.

Internet Engineering Task Force (IETF) provides its description of IoT along with definitions for "Internet" and "thing" (IETF, "Internet of Things," 2010): "The basic idea is that IoT will connect objects around us (electronic, electrical, non-electrical) to provide seamless communication and contextual services provided by them." "Development of RFID tags, sensors, actuators, mobile phones makes it possible to materialize IoT which interact and co-operate each other to make the service better and accessible anytime, from anywhere."

IETF's definition of "Internet": "The original 'Internet' is based on the TCP/IP protocol suite but any network based on the TCP/IP protocol suite cannot belong to the Internet because private networks and telecommunication networks are not part of the Internet even though they are based on the TCP/IP protocol suite. In the viewpoint of IoT, the 'Internet' consider[s] the TCP/IP suite and non-TCP/IP suite at the same time."

IETF's definition of "things": In the vision of IoT, "things" are very different such as computers, sensors, people, actuators, refrigerators, TVs, vehicles, mobile phones, clothes, food, medicines, books, etc.

These things are classified into three scopes: people, machines (e.g., sensor, actuator, etc.), and information (e.g., clothes, food, medicine, books, etc.). These "things" should be identified at least by one unique way of identification for the capability of addressing and communicating with each other and verifying their identities. Here, if the "thing" is identified, we call it the "object."

ETSI provides an architectural model for M2M communication centered on the usage of connectivity and related models. Data is also a relevant part of that architectural design. ITU describes enabler technologies that are required to bring IoT into reality. In addition to a structural model, ETSI gives a detailed design and description of protocols, addressing security for M2M communication. IEEE 2413 provides an architectural framework including descriptions of various IoT domains, definitions of IoT domain abstractions, and identification of commonalities between different IoT domains. Most of the standardization bodies emphasize the network and communication aspect of IoT but World Wide Web Consortium (W3C) works on the standardization of the Web in a way that supports IoT applications and virtual representation of IoT components on the Internet. Accordingly, merging the communication-oriented works done by other standardization bodies, like ETSI, with the software-oriented work done by W3C will allow IoT to be practical.

In the IoT architecture, the "Internet" represents the internet and its technologies and it acts as a middleware between the user and intelligent things and therefore it is called intelligent middleware. Intelligent middleware will allow the creation of a dynamic map of the real/physical world within the digital/virtual space by using a high temporal and spatial resolution and combining the characteristics of ubiquitous sensor networks and other identifiable things.

6.1 Communication Between "Things"

This is an example to explain the phenomenon once again which we have already learned in the Introduction part; when white light is shone on a red object, the dye absorbs nearly all the light except the red, which is reflected. At an abstract level, the colored surface is an interface for the object (the object called RED), and the light arriving at the object can be treated as a message sent to the thing, and accordingly, its reflection is (the color RED) the response from the thing. Such consistency in responses received from the interfaces for each message enables things to interact with their surroundings. Hence, to make the virtual world comprehensible, there is a need to be consistent in messages and their responses. In our case of IoT, this is enabled through standard interfaces, which in turn support and facilitate interoperability. Simply this phase focuses on the functionalities and communications among sensor/actuators, embedded devices, and any other smart devices.

Hence, Things may be viewed based on the context in a different way and depending on the application domain in which it is used. In industry, all IoT activities are involved in financial or commercial transactions among companies, organizations, and other entities such as manufacturing, logistics, service sector, banking, financial governmental authorities, intermediaries, etc. On the whole, the "Thing" may typically be the product itself, the equipment, transportation means, etc.—everything that participates in the product lifecycle.

Global System for Mobile (GSM) Communication is an important communication technology and one can propose "IoT Home Appliance architecture" using GSM as a primary communication technology between the home and IoT agent. The IoT agent is the core part of this proposed architecture because it manages web server data, SMS command, GSM module interactions, and all knowledge-based processes (e.g., parsing, analysis, and creation of SMS commands). One can create an interface between users and smart homes by using GSM and Internet technologies, or it simply creates GSM-based wireless communication from the web server into the smart home. In this architecture the users give commands through the web and then the users' inputs are converted into GSM-SMS commands.

Chapter 7

IoT Architecture

IEEE P2413 aims to accelerate the growth of the IoT Market by enabling cross-domain interaction and platform unification through increased system compatibility, interoperability, and functional exchangeability. Define an IoT architecture framework that covers the architectural needs of the various IoT application domains. Increase the transparency of system architectures to support system benchmarking, safety, and security assessments. Reduce industry fragmentation to create a critical mass of multi-stakeholder activities around the world and leverage the existing body of work.

This standard defines an architectural framework for the IoT, including descriptions of various IoT domains, definitions of IoT domain abstractions, and identification of commonalities between different IoT domains.

The P2413 architectural framework for IoT provides:

- Reference model that defines relationships among various IoT application domains (e.g., transportation, healthcare, etc.) and common architectural elements.
- Defines basic architectural building blocks and their ability to be integrated into multi-tiered systems.
- Addresses how to document and mitigate architectural divergence.
- Gives a blueprint for data abstraction and the quality "quadruple" trust that includes protection, security, privacy, and safety.

IEEE wishes to create an open community wherein all are welcome to participate in and share perspectives on addressing and preparing for the interconnected world of 2020, through the IoT initiative.

IoT implementations use different technical communications models, each with its specific characteristics. The Internet Architecture Board describes four common communications models: **Device-to-Device, Device-to-Cloud, Device-to-Gateway, and Back-End Data-Sharing**. These models highlight the flexibility in the ways that IoT devices can connect and provide value to the user.

DOI: 10.1201/9781003303206-7

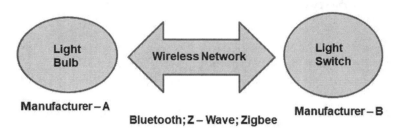

Figure 7.1 IEEE device-to-device model.

Device-to-Device: The device-to-device communication model represents two or more devices that directly connect and communicate with one another, rather than through an intermediary application server. These devices communicate over many types of networks, including IP networks or the Internet. Often, however, these devices use protocols like Bluetooth-40, Z-Wave-41, or ZigBee42 to establish direct device-to-device communications.

Figure 7.1 depicts an example of a device-to-device communication model. These device-to-device networks allow devices that adhere to a particular communication protocol to communicate and exchange messages to achieve their function. This communication model is commonly used in applications like home automation systems, which typically use small data packets of information to communicate between devices with relatively low data rate requirements. Residential IoT devices like light bulbs, light switches, thermostats, and door locks normally send small amounts of information to each other (e.g., a door lock status message or "turn on light" command) in a home automation scenario.

Device-to-Cloud: In a device-to-cloud communication model, the IoT device connects directly to an Internet cloud service like an application service provider to exchange data and control message traffic. This approach frequently takes advantage of existing communications mechanisms like traditional wired Ethernet or Wi-Fi connections to establish a connection between the device and the IP network, which ultimately connects to the cloud service.

Figure 7.2 depicts the device-to-cloud communication model. This communication model is employed by some popular consumer IoT devices like the Nest Labs *Learning Thermostat* and the Samsung *SmartTV*. In the case of the Nest Labs *Learning Thermostat*, the device transmits data to a cloud database where the data can be used to analyze home energy consumption. Further, this cloud connection enables the user to obtain remote access to their thermostat via a smartphone or Web interface, and it also supports software updates to the thermostat. Similarly, with the Samsung *SmartTV* technology, the television uses an Internet connection to transmit user

Figure 7.2 IEEE device-to-cloud model.

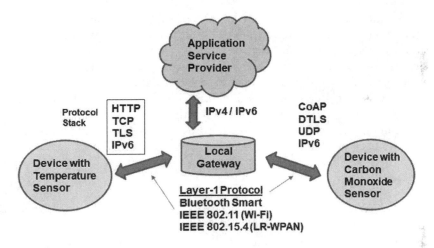

Figure 7.3 IEEE device-to-gateway model.

viewing information to Samsung for analysis and to enable the interactive voice recognition features of the TV.

Device-to-Gateway: In the device-to-gateway model, or more typically, the device-to-application-layer gateway (ALG) model, the IoT device connects through an ALG service as a conduit to reach a cloud service. In simpler terms, this means that there is application software operating on a local gateway device, which acts as an intermediary between the device and the cloud service and provides security and other functionality such as data or protocol translation.

Figure 7.3 depicts the device-to-gateway communication model. Several forms of this model are found in consumer devices. In many cases, the local gateway device is a smartphone running an app to communicate with a device and relay data to a cloud service. This is often the model employed with popular consumer items like personal fitness trackers. These devices do not have the native ability to connect directly to a cloud service, so they frequently rely on smartphone app software to serve as an intermediary gateway to connect the fitness device to the cloud.

Back-end-Data-Sharing: The back-end data-sharing model refers to a communication architecture that enables users to export and analyze smart object data from a cloud service in combination with data from other sources. This architecture supports "the user's desire for granting access to the uploaded sensor data to third parties" as an example. This approach is an extension of the single device-to-cloud communication model, which can lead to data silos where "IoT devices upload data only to a single application service provider." A back-end sharing architecture allows the data collected from single IoT device data streams to be aggregated and analyzed.

For example, a corporate user in charge of an office complex would be interested in consolidating and analyzing the energy consumption and utility data produced by all the IoT sensors and Internet-enabled utility systems on the premises. Often in the single device-to-cloud model, the data each IoT sensor or system produces sits in a stand-alone data silo. An effective back-end data-sharing architecture would allow the company to easily access and analyze the data in the cloud produced by the whole spectrum of devices in the building. Also, this kind of architecture facilitates data portability needs. Effective back-end data-sharing architectures allow users to move their data when they switch between IoT services, breaking down traditional data silo barriers.

Figure 7.4 depicts the back-end data-sharing model. This architecture model is an approach to achieve interoperability among these back-end systems. As the *IETF Journal* suggests, "Standard protocols can help but are not sufficient to eliminate data silos because common information models are needed between the vendors." In other words, this communication

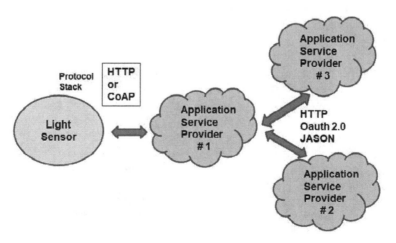

Figure 7.4 IEEE back-end data-sharing model.

model is only as effective as the underlying IoT system designs. Back-end data-sharing architectures cannot fully overcome closed system designs.

Summary: From a general user perspective, these communication models help illustrate the ability of networked devices to add value to the end-user. By enabling the user to achieve better access to an IoT device and its data, the overall value of the device is amplified. For example, in three of the four communication models, the devices ultimately connect to data analytic services in a cloud computing setting. By creating data communication conduits to the cloud, users and service providers can more readily employ data aggregation, Big Data analytics, data visualization, and predictive analytics technologies to get more value out of IoT data than what can be achieved through traditional data-silo applications. In other words, effective communication architectures are an important driver of value to the end-user by opening possibilities of using the information in new ways. It should be noted, however, that these networked benefits come with trade-offs. Careful consideration needs to be given to the incurred cost burdens placed on users to connect to cloud resources when considering one of these architectures, especially in regions where user connectivity costs are high.

7.1 Conceptual Design of IoT

This has reference to the Figure 7.5 "conceptual design architecture of the IoT".

USER: The "web user" or "user" are the common people who wish to use the smart home devices (i.e., users can make a specific home device "switch-on" or "switch-off" while from a remote location or from within the home from a different room) using a Wireless Application Protocol (WAP)-enabled device—mobile phone, tablet, PC, etc.

SERVER: Web server might be a Tomcat, Apache, IIS, etc., and it must have an internal database. User data is kept in the centralized database and it will be read by IoT Agent through Common Object Model (COM) or Distributed Common Object Model (DCOM), Application Programming Interfaces (API), or any other Open Database Connectivity (ODBC). The web server is not involved in parsing and knowledge processes. It belongs only to requests and responses to the user and stores the data. The web server has an internal timer to refresh the web page to update the status of the smart home devices.

IoT Agent: The IoT Agent is the software and hardware unit which continuously monitors the Web server and GSM embedded module for handling SMS that is aimed at a smart home. The special SMS may have a unique structure that is constructed based on the microcontroller of the GSM embedded module. Once the NEW_MESSAGE_RECEIVED event is generated by the GSM module, the IoT Agent reads the newly arrived SMS

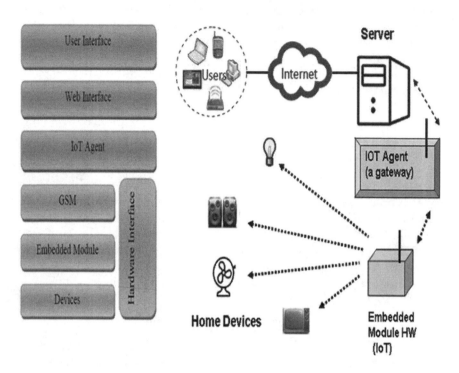

Figure 7.5 Conceptual design of IoT.

and checks it whether it is special SMS or not. If so, it parses and extracts the data.

These special SMS may contain house number (some special code tag to confirm that the house owner is only operating it), the room/hall/kitchen/code and the fan, light, or appliance identifier like refrigerator, etc., and then "switch-on" or "switch-off," etc., in a standard and systematic way.

(Note: the embedded module h/w and technology need not be limited to GSM. It can be any one of the under-laying technologies that are related to the last mile of the smart home like Wi-Fi, GSM, Zigbee, etc., which is according to the IoT framework of standards.)

The prototype models can be developed and tested with frequently used technologies. The embedded module will be placed in a typical home and attached with minimum devices. It is assumed that all the home appliances are connected with miniature Electric Sensors which can "switch-on" or "switch-off" the appliance based on a signal from the embedded hardware.

Only for convenience one can include the below GSM-based embedded hardware design for completion of the concept and discussion. Of course, many will be familiar with the microprocessor 8051 and hence it has been chosen.

Figure 7.6 depicts the conceptual design model which can be implemented on the ground also. Nowadays, there are lots of design modules available across the market wherein one can experiment and build better models.

Few of many and many IoT applications were shown here, just a birds' eye view (Figure 7.7).

At present, over three-quarters of companies are either actively exploring or using the IoT. The vast majority of business leaders believe that it will have a meaningful impact on how their companies conduct business, yet there is some divergence about the wider effect it will have. The largest group of respondents (40%) see the impact limited to certain markets or

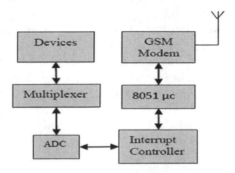

Figure 7.6 8051-based embedded hardware.

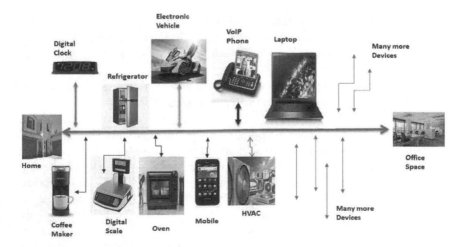

Figure 7.7 IoT applications.

industries, whereas a similar-sized group of respondents (38%) believe that the IoT will have a major impact in most markets and industries. Turning 50 billion so-called smart things into a global network requires the business to agree on standards for interconnectivity and data sharing. Service providers are already offering new IoT-based products (e.g., insurance companies' pricing premiums based on driver behavior).

In retail, businesses have problems in identifying the right customer at the right time to sell their products. Various techniques of marketing products involve using SMS broadcast, digital signages and recently the use of Quick Response (QR) codes to bundle promotions. These methods often fail to deliver the right customer to the right product and vice versa. New trends of marketing have evolved with businesses shifting from mass market advertising to context-aware systems to anticipate customer needs and proactively serve the most appropriate products or services. For example, a male shopper, looking to buy business suits for a job interview, will be informed of exact store locations selling suits that match his body size, style, and budget. Behind the scene, the context-aware system tries to understand the profile and sentiments of the male shopper and combines data from the mall to "intelligently" make recommendations to suit the shopper. Gartner has forecast that context-aware technologies will affect the US$96 billion, of annual consumer spending by 2015, with 15% of all payment card transactions being made on the back of contextual information.

RFID technology is of particular importance to IoT as one of the first industrial realizations of IoT is in the use of RFID technology to track and monitor goods in the logistics and supply chain sector. RFID frequency bands range from 125 kHz (LF) up to 5.8 GHz (SHF) and the tags have at least three basic components:

- The chip holds information about the object to which it is attached and transfers the data to the reader wirelessly via an air interface.
- The antenna allows transmission of the information to/from a reader.
- The packaging encases the chip and antenna and allows the attaching of the tag to an object for identification.

Today, coil-on-a-chip technology has been implemented in certain RFID tags and specialized applications such as magnetic resonance micro-imaging. When compared to conventional ones with external antenna coils, the coil-on-chip RFID tag achieves a smaller footprint and rarely malfunctions because of the deterioration of contacts on the lack of external soldering connections between the antenna coils and the IC chips. RFID coil-on-chip has basic storage capacities ranging from 128 bytes to 4 KB of data and no moving parts so that it can withstand the harshest environments, including wet and dry conditions.

IEEE 802.15.4 wireless technology is a short-range communication system intended to provide applications with throughput and latency requirements in Wireless Sensor Network (WSN). The key features of 802.15.4 wireless technology are short-distance transmission, low power consumption, and low-cost characteristics that can be supported by devices. Most WSNs use wireless mesh technology based on IEEE 802.15.4, sometimes referred to as ZigBee. ZigBee is a specification for a suite of high-level communication protocols using small, low-power digital radios based on an IEEE 802 standard for personal area networks (PAN). Using ZigBee protocol, sensors can communicate with each other on low-power, reliable bit-rate transfer of 250 kbps at 2.4 GHz band and secure data transfer, that is, 128 Advanced Encryption Standard (AES) plus security. The radio design used by ZigBee has been optimized for low-cost production and has a transmission range of less than 100 m.

Chapter 8

IoT for Smart City Applications

Smart city implementers today use various standards to support their governing and also citizen requirements. With numerous sources of data and heterogeneous devices, the use of standard interfaces between these diverse entities becomes important at any smart city strategy. This is especially so for applications that support cross-organizational and various system boundaries. For example, in the logistics sector, the supply chains involve multiple stakeholders like retailers, manufacturers, and logistics, hence the IoT systems need to handle a high degree of interoperability for information to be processed down the value chain, across the smart city, and also various consumers.

Smart cities should reduce resource use through optimization. The gains from optimization and improved planning mean that cities, their businesses, and their residents consume less water, gas, and power. Smart cities also reduce the duplication of effort and reduce costs through infrastructure sharing. Here are a few of the elements that can often be purchased or designed just once and reused many times: geographic information systems (GIS), communication networks, cybersecurity designs and implementations, database management systems, enterprise service buses, workforce and field crew management architecture, operations centers that can be interconnected through standards-based technology implementations. IoT-enabled utility and logistics, etc., which can bring in major changes and improved living styles in smart cities.

Smart cities are helping improve their citizens' lives by opening up data for application development to disseminate timely information about public safety, public health, transportation, and other services that impact the public. For instance, the INRIX Traffic application that can be used on a Windows Phone helps users decide which route is the best choice to get around traffic.

Developing a methodology for "customer profiling" for use by local municipalities will help municipalities target their services by having a better understanding of the needs of local citizens. One can develop and

DOI: 10.1201/9781003303206-8

implement a range of customer profiling methods, including online survey tools, and marketing training.

Key groups one should include for customer profiling are the following:

Families that have high levels of contact with different arms of public bodies because they have problems of health, social care, housing, education, money, crime and employment, and people who have low skills.

Every smart city should implement a geo-coded system tool to facilitate the provision of "geo-located services", and its connected terminals at most of the important government offices, tourist centers, and local train terminals, and make them freely available to citizens and visitors, and improve its specific processes as the new e-services are developed. These will be the first steps for any city that wishes to develop as a smart city, and develop a basic and serviceable e-government environment, and act as the focal point for the city's efforts to become a more user-friendly, efficient, and modern city for citizens and visitors.

Regular workshops on smart cities themes, bringing a wide range of knowledge and best practices from the smart cities network to the required region, including service architectures, customer profiling, online surveys, customer contact centers, personalized online services, web services, GIS, and innovation.

The web platform used by the local city region municipalities will be enhanced to enable municipalities to provide **authenticated and secure transactions** (Aadhar card), personalized information, and geo-based services. A **regional contact database**—developed in cooperation with the city—will lead to better contact data as a basis for customer relationship management in the municipalities.

Smart cities should develop pilots and methods for **involving users in e-service development** that is based on what one has learned. Smart city-governed websites should be created and invite the public to identify service needs that have led to the development of pilots on child care, press information, and other location services like road traffic, parking spaces, etc.

The Business Process Change application will incorporate both lean thinking and customer journey mapping approaches into the smart city strategy to improve their internal business processes. The approach should be linked with the smart city's Revenues and Benefits Division. In addition to aiming to identify measurable process improvements, the pilot can develop training and tools that will be used across the nation to support business process change, that will be used across the new businesses. This will allow one to deliver other customer service process reviews and to share their methodologies with smart cities partners.

The procurement and development of a new Internet presence for the smart city will give the capability to deliver a more useful, accessible, and usable website with up-to-date, reliable, and accurate information. The ability to complete transactions online requires a new platform that is both adaptable and scalable to incorporate future innovations and new technologies, for smart city dwellers.

Wireless service pilot shall give the provision of free wireless Internet access for citizens in some of the city's local libraries. The lessons learned from this experience will be shared with all smart cities partners and will provide valuable insights for any further initiatives to provide wireless services—be it in libraries or elsewhere.

Smart-office—for a smart city municipality to appear as a single entity to the public, companies, or visitors, we will need systems that seamlessly work across the various parts of the municipality. It is not enough to create a variety of projects: a framework is required with a common strategy, methodology, technology, and knowledge to succeed local municipality that has created an organization called the smart-office, which brings together different models, guidelines, templates, approaches, and common solutions. Much of this knowledge can be developed with the help of regional universities and with input from other project partners. The main task of smart-office is to coordinate, to help departments to understand the benefits of electronic service and e-governance, to provide required modifications to existing e-services, and to support the development of e-services for smart cities as an ongoing change.

My Page—One of the municipalities' goals in the smart cities project is to introduce the "my Page" approach, where municipal web pages are personalized for citizens. City management can put together a roadmap for the sustainable development of such services, which will be used to get a broader overview of what is required and to set out our approach to the development of personalized web pages. An action plan and roadmap should be in place for each smart city.

Portal for Wi-Fi networks and I-points—The portal that provides content for the I-point digital tourism kiosks will be accessible via the smart city's Wi-Fi networks and on the university campuses. It will collect information from a range of existing databases and deliver this to touch screens, mobile devices, and normal computers.

Integrated solutions link municipal invoicing data with accounts receivables, and produce digital invoices that are sent directly to the relevant banks within the smart city or even outside in a different city. The municipalities are also developing automated processes, where services are automatically provided if the user submits the correct data (e.g., booking an appointment).

It is to be noted that herein we are not discussing the solutions related to the infrastructure streams like different transports such as local metro/tube trains, roadways, electricity, water supply, etc., without which one may not call the city as a smart city.

8.1 Indian Government and IoT and Smart City Technology Prospective

The Government of India has allocated 70.6 billion (US$1.2 billion) for smart cities in Budget 2014–15. Note that these indications are estimates in 2014–15, that may change from time to time, but gives birds-eye view on Indian Government's interest about smart cities. Public Private Participation (PPP) model are to be used to upgrade infrastructure in 500 urban areas. In order for the smart city projects to create a 10%–15% rise in employment, the Ministry of Urban Development has plans to develop two smart cities in each state of India. Delhi Mumbai Industrial Corridor Development Corporation Ltd (DMICDC) plans seven "smart cities" along the 1,500 km industrial corridor across six states with a total investment of US$100 billion.

India was expected to emerge as the one of the world's largest construction markets soon, by adding 11.5 million homes every year (refer Figure 8.1).

The Intelligent Building Management Systems market was around US$621 million and was expected to cross US$1,891 million by 2016.

Smart Buildings estimated to save up to 30% of water usage, 40% of energy usage, and a reduction of building maintenance costs by 10%–30%.

Smart Grid

> Electrification of all households with power available for at least 8 hours per day by 2017.
> Indigenous low-cost smart meter by 2014.
> Establish a smart grid test bed by 2014 and a smart grid knowledge center by 2015.
> Implementation of eight smart grid pilot projects in India with an investment of US$10 million.

Figure 8.1 Smart city.

Energy Storage

Addition of 88,000 MW of power generation capacity in the 12th Five Year Plan (2012–17).
India needs to add at least 250–400 GW of new power generation capacity by 2030.
The Power Grid Corporation of India Ltd has planned to invest US$26 billion in the next five years.

Smart Meters

India to install 130 million smart meters by 2021.

Renewable Energy

The Ministry of New and Renewable Energy has plans to add a capacity of 30,000 MW in the 12th Five Year Plan (2012–17).

Water and Waste Water Management

The Indian Ministry of Water Resources plans to invest US$50 billion in the water sector in the coming years.
The Yamuna Action Plan Phase III project for Delhi is approved at an estimated cost of US$276 million.

Sanitation

About 67% of the rural population continues to defecate in the open, and India accounts for about 50% of the world's open defecation.
The Government of India and the World Bank have signed a US$500 million credit for the Rural Water Supply and Sanitation (RWSS) project in the Indian states of Assam, Bihar, Jharkhand, and Uttar Pradesh.

8.1.1 Lavasa Dream

Great dreams are realized only by planning. Lavasa, a planned hill station in Maharashtra, is one such project. Aimed as India's first hill city since independence, it is developed primarily by Hindustan Construction Company (HCC) India and is set amid seven hills and 60 km of lakefront and spread over 25,000 acres. It is a convenient 3 hours' drive from Mumbai, 1 hour's drive from Pune, and is a whopping quarter size of Mumbai. Lavasa is planned across four town centers. Lavasa city will have a wide range of residences, from sprawling hillside villas to up to 3-bed, hall, and kitchen (3 BHK) homes, and will offer homes that fit budgets across socio-economic classes. It is expected to provide abundant opportunities as a global leader in hospitality, health, wellness, and education. Lavasa Tourism will be setting up its institutions. In addition, this hill station in Maharashtra has diverse work possibilities appealing to the IT and biotech industry, Knowledge Process Outsourcing firms (KPOs) and Research and Development (R&D) companies, and the world of art, fashion, and animation. One of the largest private infrastructure projects in India, Lavasa city is planned for a permanent population of 0.2 million residents and a Lavasa Tourism inflow envisaged at 2 million per annum.

8.1.2 The Smart Cities in India

According to the report on "India's Urban Awakening" by McKinsey Global Institute, in the next 20 years, India will have 68 cities with a population of over 1 million—up from 42 today. That is near twice as many cities as all of Europe. Most cities in Europe and America were established in the 19th century when there was easy availability of land, gas, and water. India is a late starter and is far more crowded and complex. Therefore, India requires a far more efficient and sustainable solution for servicing urban areas and can reap the benefits by using technology to learn from practices from other parts of the world. Thus, India, too, is on the road to building smart cities—world-class, self-sustainable habitats with minimal pollution levels, maximum recycling, optimized energy supplies, and efficient public transportation. The cities would come along Delhi Mumbai Dedicated Rail Freight Corridor which is under implementation. In this endeavor to transform the rapidly growing urban areas into smarter cities, a collaborative partnership between government, industry, academia, and civil society will the pave way for the attainment of this dream.

- Smart parking: Monitoring of parking spaces available in the city.
- Structural Health: Monitoring of vibrations and material conditions in buildings, bridges, and historical monuments.
- Noise Urban Maps: Sound monitoring in bar areas and centric zones in real time.
- Smartphone detection: Detect smartphones and in general any device that works with Wi-Fi or Bluetooth interfaces.
- Electromagnetic field levels: Measurement of the energy radiated by cell stations and Wi-Fi routers.
- Traffic congestion: Monitoring of vehicles and pedestrian levels to optimize driving and walking routes.

- Smart lighting: Intelligent and weather adaptive lighting in street lights.
- Waste management: Detection of rubbish levels in containers to optimize the trash collection routes.
- Smart roads: Intelligent highways with warning messages and diversions according to climate conditions and unexpected events like accidents or traffic jams.

Chapter 9

IoT Middleware

IoT defines a new method and delivers a paradigm shift to business view, where the real, the digital, and the virtual worlds converge to create an environment that makes the energy, transport, city, and many other areas more intelligent and easily accessible to manage. The IoT is a concept and includes various environments (energy, transport, city, building, etc.) and a variety of things/objects through wired or wireless that are uniquely addressed and can interact with each other and cooperate with other things/objects to create new methods of applications/services and to achieve common objectives. The purpose of IoT is to validate the connection type: anytime, anywhere, and everything and everyone.

In previous chapters, we have learned about the sensors and 6LoWPAN (PHY, MAC) and Network. Now let us concentrate on the middleware aspects of IoT architecture. IoT can be divided into a layered architecture designed to answer the demands of various industries, enterprises, and societies. Figure 9.1 shows a generic layered architecture for IoT that consists of five layers, which are discussed in the following. Also note that as we are representing architecture, this may differ from the true representation of International Standards Organization (ISO) seven layers but is practical to the designer's understanding.

Edge: At the hardware level there shall be ongoing research and innovation for the development of different miniature sensors, actuators, and other devices including RFID, for the next few years as the IoT systems can identify each of the unique required inputs and collect the information in a heterogeneous environment.

Access: IoT network may be configured using mobile, wireless, or wired environments and should support bidirectional reliable communication at different levels like application and services with highly distributed structure while supporting low energy, long battery life, etc. The introduction of the IETF 6LoWPAN family of protocols has an essential role in connecting low-power radio devices to the Internet (Network Layer) and the working group from IETF ROLL introduced appropriate routing protocols to achieve universal connectivity.

DOI: 10.1201/9781003303206-9

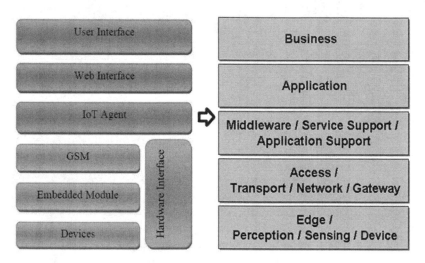

Figure 9.1 IoT layered architecture.

Middleware: IoT networks are made of heterogeneous devices to collect different specific inputs regularly. Hence, the supporting middleware should shield their heterogeneity and seamlessly communicate with Services and Application levels by the way of delivering different required information packets which shall enable the security and scalability of IoT architecture.

Application/Business: Based on the application and business requirements the gathered inputs shall be judged and evaluated by the respective application and business ware to enable the management with the right inferences on the management.

At this juncture, we are more concentrating on the middleware which is more useful for the design of the IoT devices. In this discussion, we deal with more concepts rather than the steps leading to final functions and code combinations.

9.1 Middleware Layer

The middleware layer is also called the processing layer because it stores, analyzes, and processes the information related to items received from lower layers. This is the layer where the IoT systems run. To modularize the physical objects, a proxy can map the messages of the objects to their logical components from the middleware.

Technology experts like Bandyopadhyay et al. have studied the middleware systems that have been applied in IoT-based systems. They classify the required functionality of middleware to manage interaction with a variety

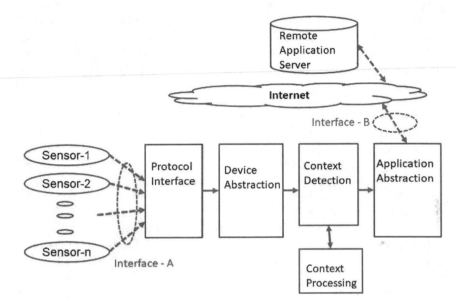

Figure 9.2 IoT middleware functions.

of devices into four functional components, namely (1) interface protocols, (2) device abstraction (DA), (3) central control, context detection, and management (CCM), and (4) application abstraction (shown in Figure 9.2). In the following, we explain these components in detail.

Interface Protocols: Interface protocol defines the ability of two systems to interoperate by using the same communication protocols. According to ETSI, technical interoperability is defined as the association of hardware or software components, systems, and platforms that enable M2M communication to take place. This kind of interoperability is often centered on (communication) protocols and the infrastructure needed for those protocols to operate.

Device Abstraction (DA): DA provides syntactic and semantic interoperability which creates formats to facilitate the interaction of the application components with devices. DA components provide two general functionalities: (1) to ask devices to perform some functionality and (2) to define and configure devices.

Syntactic interoperability is associated with data abstraction part (or data formats). The messages transferred by communication protocols must have a well-defined syntax and encoding format, which can be represented by using high-level transfer syntaxes such as Hyper Text Markup Language (HTML) and Extensible Markup Language (XML).

Semantic interoperability is also a part of data abstraction, but it is related to the content code, the meaning of the content of the message which is understandable for humans. Thus, interoperability on this level means that there is a common understanding among people/or all the objects/or things involved, in the meaning of the content (information) being exchanged among them. Semantic interoperability relies on semantic models which tend to be domain-specific. For example, one way to provide semantic interoperability in Service-Oriented Architecture (SOA)-based middleware is by using Devices Profile for Web Services (DPWS), and in this context, each device type refers to a distinguished service type.

Central Control, Context Detection, and Management (CCM): Context characterizes the situation of an entity, which can be a place, a person, or an object that is relevant to the user, applications, and their interactions. CCM functional component is responsible to support context-aware computation that is a computational style that takes into account the context of the entities that interact with the system. A middleware for IoT-based systems must be context-aware to work in smart environments. Smart environments refer to a physical world that is richly and invisibly interwoven with sensors, actuators, displays and computational elements, embedded seamlessly in the everyday objects of our lives, and connected through a continuous network.

Context-awareness includes two functionalities:

- **Context detection, which consists of collecting data from resources and selecting the information that can have an impact on the computation.**
- **Context processing, which is to use gathered information to perform a task or make a decision.**

Application Abstraction (AA): AA component provides an interface for both high-level applications and end-users to interact with devices. For instance, this interface can be a RESTful interface or can be implemented with some query-based language. The purpose of RESTful interface is to facilitate the interaction of high-level applications with sensors, which can communicate with WSN through the HTTP operations: (1) GET to issue a query on an existing resource, (2) DELETE to remove an existing resource, (3) POST to create a new resource, and (4) PUT to update an existing resource.

For instance, a client application gets the result of a domain task by sending a GET request through the

URL: "http://{hostname}/REST/{version}/DomaintaskResult/ id"
[In this URI that reference to an "id" leads to a unique result.]

Review of Different Middleware for IoT: Let us examine the different IoT middleware based on various features like interoperation, device management, platform portability, context awareness, security and privacy, and the support of various interface protocols. Tables 9.1 and 9.2 present the classifications of various IoT middleware systems based on the various features and interface protocol support, respectively.

Since the vision of IoT is almost similar to Pervasive computing, let us review the kinds of literature that address both IoT and Pervasive systems to select the middleware to discuss. For example, let us take HYDRA, because it is the most popular and well-documented middleware in the IoT area, compared to the below-mentioned middleware.

All the below-listed middleware(s) support device discovery and management. Context-aware functionality is supported by HYDRA and UBIROAD

Table 9.1 IoT middleware features

	IoT Middleware Features				
IoT middle-ware	Device management	Interoperation	Platform portability	Context awareness	Security & portability
HYDRA	Y	Y	Y	Y	Y
SIRENA	Y	Y	Y	N	Y
SOCRADES	Y	Y	Y	N	Y
ISMB	Y	N	Y	N	N
WHEREX	Y	Y	Y	N	N
UBISOAP	Y	Y	Y	N	N
UBIROAD	Y	Y	Y	Y	Y
GSN	Y	N	Y	N	Y

Y = Yes; N = No

Table 9.2 IoT middleware interface protocol support

	IoT middle-ware interface protocols				
IoT Middle-ware	ZIGBEE	RFID	WIFI	BLUETOOTH	Sensor
HYDRA	Y	Y	Y	Y	Y
SIRENA	N	Y	N	Y	Y
SOCRADES	N	Y	N	N	Y
ISMB	N	Y	N	N	Y
WHEREX	Y	Y	Y	Y	Y
UBISOAP	N	Y	Y	N	Y
UBIROAD	N	Y	Y	Y	Y
GSN	N	Y	Y	N	Y

Y = Yes; N = No

only. On the other hand, SOCRADES, GSN, UBIROAD, and HYDRA are some examples of middleware implementing security and user privacy in their architecture. Based on platform portability, syntactic resolution,— HYDRA is OSGi compliant; UBIROAD uses JAVA and XML; UBISOAP uses J2SE and J2ME; GSN uses XML and SQL; SIRENA and SOCRADES use DPWS; SOCRADES uses SAP NetWeaver platform also; ISMB uses any JAVA compliant platform. WHEREX is developed using J2EE architecture and is integrated with Oracle Application Server. It also uses the Rhino rule engine which is an implementation of JavaScript.

Some of the middleware(s) have shortcomings or open issues, which might have been corrected or updated, that are to be checked by the designers before they take a call to use. Some are available for specific domains separately. Most of them address the RFID domain. GSN addresses the sensor networks in general. UBIROAD addresses smart vehicular systems. There exists no generic middleware that can be applied across all possible smart environments—like smart home, smart vehicle, smart city, etc., including RFID domain, and can be customized as per the domain-specific requirements, as of now.

Chapter 10

The Core Value of IoT Devices and Design Propositions

Designers need to learn about the core value proposition of any IoT device they are planning to design. Customers need not bother about the IoT device connectivity or its on-board computing, etc. The core value proposition of any IoT device should be straightforward. The user should be able to easily know about the utility value of the IoT device, its environment—whether it is meant for outdoor or indoor operations, its reliability, its life span, interoperability, and most importantly what are the focused results that it can deliver during its operational lifetime.

The challenge is to convince the users that this specific IoT product is the best solution. One should be able to explain how the other competitive systems can even interoperate with the specific IoT device and the additional benefits it can deliver. Note that users do not care for the technology you use for design, but they look for the reliability of the IoT device at hand.

The designer needs to keep in mind the following four different markets and their needs before designing an IoT device:

(a) **Low-Cost Product in the Existing Market:** There shall be always a guy who produces a product which is a bit cheap and hence users are attracted. But it is the reliability and stability of the product that always win; hence, designers should not sacrifice reliability, stability, interoperability, etc., while designing a product. For example, people buy street-side watches for their kids or even for themselves for about one or two hundred rupees, but the value of any branded watch has not come down because of this.

(b) **Niche Product in the Existing Market:** Always niche products will have their share in the existing markets. But remember IoT devices are meant for mass utility, and they are designed to sell in thousands and maybe millions. Designers need to add a niche value (additional feature) into the design of the product which can convince the mass markets in a real quick time. This should differentiate the other similar products from this specific product in the market.

(c) **New Type of Product in the Existing Market:** This is a continuous game and designers always have to come out with a new product to address the market to keep their brand image in the existing markets. Designers

DOI: 10.1201/9781003303206-10

should keep in mind that they cannot fool the customers with old wine in a new bottle, as in such cases markets lose momentum very quickly.

(d) **New Product and New Market:** New intelligence and additional features with embedded functionality shall fuel new markets and new products. It is indeed a continuous process of evaluation. Here the challenge is to convince the users about the utility value of the new product compared to the available niche products in the market.

Each designer must need to learn a lesson from the Nokia experience, which once ruled the global markets with their reliable, stable, and nice-looking cell phone devices. Nokia missed the boat of the smart cell phone devices by one or two years and slowly they lost their value proposition in the cell phone markets to Apple, Samsung, etc. Soon Nokia may be read in the history, or how Nokia shall make a comeback with new vigor is to be seen.

It is to be noted by designers that mass market consumers, who do not have deep technical or domain knowledge, generally expect a product to come designed and engineered to fulfill a specific need. The Nest Protect smoke detector and carbon monoxide alarm is a good example of a product. The marketing website focuses on how it has a better safety alarm, etc. Connectivity is only mentioned at the end, to say you'll be alerted on your phone if there's a problem when you're away from home.

A communicated value proposition for any product (in our case IoT device) is fundamental to user experience. When people come across a product (or service), they try to form a quick judgment about its purpose, and whom it is for. If it's not immediately clear what the value proposition is, it may be dismissed: either because it is too hard to figure out or because it does not appear to do anything of value for that person at that time. Worse, potential users may wrongly assume it can fulfill a purpose for which it is not suited and waste time and/or money on a fruitless endeavor. (You may be happy to take their money in the short term, but over time too many unhappy customers will damage your reputation!).

It is needless to say that the target system needs to be reliable enough to fulfill its promise. Glitches and outages are inevitable in most systems and early adopters will forgive these more readily. But if there are contexts of use in which you cannot afford failure, the product must be 100% reliable. For example, emergency alarms for elderly or vulnerable people must always work. You'll need a backup power supply and connectivity and regular checks to ensure their reliability.

Imagine that you're designing a wrist-top device for outdoor pursuits like hiking or climbing. The core features are an altimeter, barometer, compass, and perhaps GPS. It might be quite straightforward software-wise to add on a calendar, to-do list, and, maybe, games. You can probably imagine a situation in which someone, somewhere, might use those features. But you'll be at risk of obscuring the key purpose of the device: helping users find their way and stay safe. Too much flippant functionality might even undermine the perception that the device offers good quality in its core functionality. And it will make it harder for users to access the key features they most want and need.

A **security alarm** is an example of a system where the *service* is the focus: we might call it a *device-enabled service*. The alarm service is what users care about. The sensors and other devices are generally low profile and most of the intelligence sits in the Internet service or gateway software. You could add or swap out devices without affecting the core functions of the service (e.g., Alexa). Key factors that indicate that **service** may be the focus of your user experience, not the device itself. Interactions are distributed across multiple devices, so no single device is the center of attention.

Most functionality lives in the cloud service or gateway software (perhaps because local devices don't have much computing power); and/or

Devices can be added, removed, or swapped without changing the core functioning of the system.

Over time, as we all become more accustomed to the IoT products around us having intelligence and connectivity, our ability to understand connected products as *services* without depending on physical manifestations may become more sophisticated. The idea of a heating system without a visible controller or a door lock without a visible lock may seem strange, but in time, as long as they work, we (customers) might be more open to such things.

Note that each IoT is part of a service ecosystem. Services are delivered through the interactions of networks of people, organizations, infrastructure, and physical components. The devices, and even the digital components, are only part of the experience. As a part of the Delhi Metro experience, one may consider the interactions one may have with Metro staff when buying a ticket/smart pass or asking for help, etc. To help you, Metro staff may have been trained to provide good customer service, but they will also need access to good information about the transactions on your card and system information. Making this whole system work smoothly is a lot more complicated than just making cards, tickets, machines, and a website: it requires someone (the designer) to take a holistic view of how the service is experienced, and make sure that all the components work reasonably smoothly and together.

Each of the IoT products might be focused around the device or the service. All IoT systems depend on some kind of digital service, and perhaps offline service components too, like professional installation, maintenance, or customer support helplines. Ensuring this work well together is an important part of the overall design responsibility and designers have to take care of this.

Below are some of the IoT designers' paradigms:

- What best fits the situation and context?
- What is the right composition?
- Can my device work with pre-existing devices?

- Does the system need to work if some devices are unavailable?
- How accurately does sensing need to be?
- Do users have set expectations from these devices or services?
- How do you balance cost, upgradeability, and flexibility?
- What connectivity and power issues do one need to consider?
- Consistency?
- Data and content synchronization?

10.1 IoT Application Use Cases

To address the IoT designer paradigms, here we are providing a few application areas where the respective market domains may get benefited from the LoWPAN design models. Designers should note that these may not be the complete possible list but may give some useful guidelines.

Basic Requirements Common for all Application Areas: Requirements of Low power and Lossy Networks (LLN) are similar to LoWPANs and hence the below utility functions define them (Table 10.1):

Network Nodes: LoWPAN or one may call it IEEE 802.15.4 Standard distinguishes between two types of nodes: **reduced function devices (RFDs)** and **full-function devices (FFDs)**. 6LoWPANs can be deployed using either route-over or mesh-under architectures. As the choice of route-over or mesh-under does not affect the applicability of 6LoWPAN technologies, we will use the term "6LoWPAN" to mean either a route-over or mesh-under network sometimes or vice versa.

Some may call it a **"Logical Controller"** (LC), or Local Controller, the entity that performs the special role of coordinating and controlling its child nodes for local data aggregation similar to that of RFD. And the other entity is called as **"LoWPAN Border Router"** (LBR), or Local Boarder Router, which holds the responsibility for IPv6 prefix propagation, and also works like a junction, which separates LoWPAN with other upper layers of the network. LBR can be similar to the FFD in our discussion. The intermediate LoWPAN nodes (LC or RFD) act as packet forwarders on the link layer or as LoWPAN routers and connect the entire LoWPAN in a multi-hop fashion. LBRs or FFD are used to interconnect a LoWPAN to other networks or to

Table 10.1 Low-power lossy network (LLN) requirements

Utility	Functionality	Clock/rate/speed/type
Low-cost processors	8 bit, 16 bit, 32-bit processors	10MHz–20MHz (ex: ARM7)
Less Memory	64 Kb, 512 Kb, 1Mb	EEPROM, RAM, FLASH
Low Power	10 to 30 mA	10–100 meters Range
Low Range	10 to 100 meters	0 to 3 dBm power
Low Bit Rate	20, 40, 100, 250 Kbits/s Max	IEEE 802.15.4 standard

form an extended LoWPAN by connecting multiple LoWPAN networks. Before LoWPAN nodes obtain their IPv6 addresses and the network is configured, each LoWPAN executes a link-layer configuration according to the specified mechanisms by using a coordinator that is responsible for link-layer short address allocation.

To summarize, there are three types of nodes: the first type are simple LoWPAN nodes which one may call a child (node) that will talk to its LC or LBR depending on the configuration. The second type is the LC, which controls its group of LoWPAN nodes in a star or mesh mode and forwards the required data to the LBR. The third type is the LBR itself which works as a junction and demarks between different LoWPAN networks or demarks between IP network and LoWPAN network, depending on the configuration and its position.

Network Size: The network size takes into account nodes that provide the intended network capability. The number of nodes involved in a LoWPAN could be small (ten), moderate (several hundred), or large (over a thousand).

Network Deployment: LoWPAN nodes can be scattered randomly, or they may be deployed in an organized manner in a LoWPAN. The deployment can occur at once, or as an iterative process. The selected type of deployment has an impact on node density and location.

Network Power Source: The power source of nodes, whether the nodes are battery-powered or mains-powered, influences the network design. The power may also be harvested from solar cells or other sources of energy. Hybrid solutions are possible where only part of the network is mains-powered. At present 10-year supporting long-life batteries are also are on the board for designers.

Connectivity: Nodes within a LoWPAN are considered "always connected" when there is a network connection between any two given nodes. However, due to external factors (e.g., extreme environment and mobility) or programmed disconnections (e.g., sleeping mode), network connectivity can be from "intermittent" (i.e., regular disconnections) to "sporadic" (i.e., almost always disconnected). Differences in L2 duty-cycling settings may additionally impact connectivity due to highly varying bit rates.

Security Level: LoWPANs may carry sensitive information and require high-level security support where the availability, integrity, and confidentiality of the information are crucial.

10.2 IoT Application Scenarios

There are many application areas and industrial monitoring involves Process Monitoring and control, Machine Surveillance, Storage Monitoring, Supply Chine Management, Sales Force Management, etc. Structural Monitoring involves Bridge Safety and Surveillance, Building and Room Management, etc. Smart Homes or Connected Homes require home automation and management, and health care requires patient health care at home, hospital patient care, and hospital instrument care, to name a few.

Note that the target functionality is completely different for each of the above scenarios. For example, the requirements of hospital patient care are different from that of hospital instrument care. In the same hospital room, there may be patients on treatment and also some of the medical instruments. The required IoT devices for patient care which may include some pulse monitors, etc., will talk to their respective LoWPAN network, and the other IoT nodes that monitor the medical instruments in the same room will talk to a different LoWPAN network, thus acting for totally different purposes.

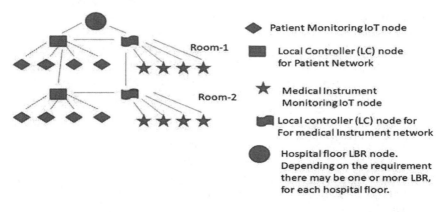

Figure 10.1 Hospital floor IoT network scenario.

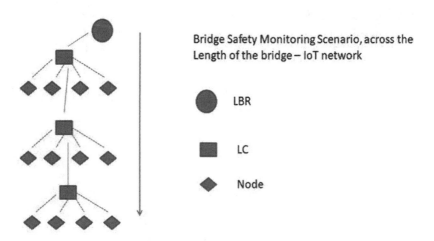

Figure 10.2 Bridge safety surveillance IoT network.

In the above hospital LoWPAN (Figure 10.1) to which the patient's information is transferred needs to operate an additional identification system, together with a strong authority and authentication mechanism. The connection between the LBR at home and the LBR at the hospital must be reliable and secure, as the data is privacy-critical. To achieve this, an additional policy for security between the other LoWPANs is recommended. Note that at hospital LoWPAN may mix up with the wired and wireless nodes, depending on how the security arrangements and management are made.

Above Bridge Safety network (Figure 10.2) shall operate in a different environment, some of the IoT nodes may monitor the local atmosphere—temperature, humidity, etc., while one of the IoT nodes may monitor the traffic on the bridge, etc., depending on the configuration and requirements.

Chapter 11

IoT Design Approach

LoWPAN-related devices should conform to the IEEE 802.15.4-2003 standard by the IEEE [IEEE802.15.4] and are characterized by short range, low bit rate, low power, and low cost.

The following are few important characteristics that need to be taken care of:

- The maximum physical layer packet is 127 bytes, thus resulting in the maximum frame size at the media access control layer being 102 octets. Link-layer security imposes further overhead, which in the maximum case (21 octets of overhead in the AES-CCM-128 case, while 9 and 13 for AES-CCM-32 and AES-CCM-64, respectively) leaves about 81 octets for data packets.
- Support for both 16-bit short or IEEE 64-bit extended media access control addresses.
- Low bandwidth. Data rates of 250 kbps, 40 kbps, and 20 kbps for each of the currently defined physical layers (2.4 GHz, 915 MHz, and 868 MHz, respectively).
- Topologies include star and mesh operation.
- A very large number of devices are expected to be deployed during the lifetime of the technology.
- The location of the devices is typically not predefined, as they tend to be deployed in an ad hoc fashion. Furthermore, sometimes the location of these devices may not be easily accessible. Additionally, these devices may move to new locations.
- Devices (hosts or nodes) within LoWPAN tend to be unreliable due to a variety of reasons: uncertain radio connectivity, battery drain, device lockups, physical tampering, etc.
- In many environments, devices connected to a LoWPAN may sleep for long periods to conserve energy, and are unable to communicate during these sleep periods.

DOI: 10.1201/9781003303206-11

As we are planning to use the IP technology, this may give some benefits or one may say limitations. IP-based technologies already exist, are well-known, and have proven to be working. Tools for diagnostics, management, and commissioning of IP networks already exist. IP-based devices can be connected readily to other IP-based networks, without the need for intermediate entities like translation gateways or proxies. A large number of devices pose the need for a large number of addresses that are readily met by IPv6.

Mesh topologies imply multi-hop routing to the desired destination. In this case, intermediate devices act as packet forwarders at the link layer (akin to routers at the network layer). Typically, these are "full-function devices" (FFD) that have more capabilities in terms of power, computation, etc., compared to the RFD, and both will exist in the network.

Applications within LoWPANs are expected to originate small packets, including all layers for IP connectivity, and also should still allow transmission in one frame, without incurring excessive fragmentation and reassembly. Furthermore, protocols must be designed or chosen so that the individual "control/protocol packets" fit within a single 802.15.4 frame. Along these lines, IPv6's requirement of sub-IP reassembly may pose challenges for low-end LoWPAN devices that do not have enough RAM or storage for a 1280-octet packet.

Accordingly, protocols used in LoWPANs should have minimal configuration, preferably work "out of the box", be easy to bootstrap, and enable the network to self-heal given the inherent unreliable characteristic of these devices. The size constraints of the link-layer protocol should also be considered. Network management should have little overhead, yet be powerful enough to control the dense deployment of devices.

IEEE 802.15.4 mandates link-layer security based on Advanced Encryption Standard (AES), but it omits any details about topics like bootstrapping, key management, and security at higher layers. Of course, a complete security solution for LoWPAN devices must consider application needs very carefully.

Security: Security of LoWPANs is equally important and one needs to take necessary care while designing itself. One argument for using link-layer security is that most IEEE 802.15.4 devices already have support for AES link-layer security. AES is a block cipher operating on blocks of fixed length, i.e., 128 bits. To encrypt longer messages, several modes of operation may be used. The earliest modes described, such as ECB, CBC, OFB, and CFB, provide only confidentiality, and this does not ensure message integrity. Other modes have been designed which ensure both confidentiality and message integrity, such as CCM* mode. 6LoWPAN networks can operate in any of the previous modes, but it is desirable to utilize the most secure modes available for link-layer security (e.g., CCM*), and build upon it. "Counter with CBC-MAC" (CCM) is referred to in RFC310, and a generic authenticate-and-encrypt block cipher mode. CCM is only defined for use with 128-bit block ciphers, such as AES. The CCM ideas can easily be extended to other block sizes, but this will require further definitions.

All implementations shall limit the total amount of data that is encrypted with a single key. The sender shall ensure that the total number of block cipher encryption operations in the CBC-MAC and encryption together shall not exceed 261. (This allows nearly 264 octets to be encrypted and authenticated using CCM, which should be more than enough for most applications.) Receivers that do not expect to decrypt the same message twice may also check this limit. The recipient shall verify the CBC-MAC before releasing any information such as the plaintext. If the CBC-MAC verification fails, the receiver shall destroy all information, except for the fact that the CBC-MAC verification failed.

11.1 IoT and 6LoWPAN—Frame Design

In the future, wireless sensor networks, or we can safely say IoT sensor networks, have to represent "billions of information devices embedded in the physical world"; hence, it is recommended to use the standard internetworking protocol. In this note, we wish to guide IoT designers about their understanding of the IPv6 datagram formats and the ease of implementation of the standards-based transmission of IPv6 datagrams on IEEE 802.15.4 networks—RFC6282.

Wireless sensor networks are based on miniature hardware with sensors, radio, and batteries. These are spatially distributed and possibly mobile with limited power and around 250 kbps communication bandwidth. These devices are aimed at long-term large-area deployments for unattended operations.

IEEE 802.15.4 corresponds to MAC and PHY layers (Layer 1—PHY and Layer 2—MAC) only. For example and clarity, ZigBee corresponds to Layer 3 – Network. One can use either ZigBee or IP or any other network protocol (at Layer 3) and can use the IEEE 802.15.4 at the lower level and transmit and receive packets. For low-power sensor networks or for IoT networks, "IPv6 over IEEE 802.15.4" (6LoWPAN) shall be the best case for the transmission of datagrams. Note that even though the standard defines many modes, some optimization shall be achieved always by the way of using only a set of standards-based IEEE frame formats to make the transmission easy and secure (Figure 11.1).

Through this note, we wish to enhance the understanding of datagram formats, LOWPAN_IPHC, for effective compression of unique local, global, and multicast IPv6 addresses based on the shared state within contexts. Besides, the standard (RFC6282), which introduces some additional improvements over the header compression format defined in RFC4944, may be checked for possible use. LOWPAN_NHC, the standard defines

N= variable octets Pay Load.		Maximum Length = 25 octets.			IEEE 802.15.4 Frame						
Octets	2	1	0/2	0/2/8	0/2	0/2/8	N	2			
	Frame Control	Sequence Number	Destination PAN Identifier	Destination Address	Source PAN Identifier	Source Address	Pay Load Variable	Frame Check			
Bits		0,1,2	3	4	5	6	7,8,9	10,11	12,13	14,15	
		Frame Type	Security Enabled	Frame Pending	Ack. Requested	Intra PAN	Reserved	Destination Address Mode	Reserved	Source Address Mode	

Figure 11.1 IEEE 802.15.4 frame detail.

a compression mechanism for User Datagram Protocol (UDP) as well as encoding formats for IPv6-in-IPv6 encapsulation, and IPv6 Extension Headers.

Important factors while sending IPv6 datagrams on IEEE 802.15.4 are header length, fragmentation requirements, routing under IP topology, routing over mesh 802.15.4 nodes, and low energy calculations [per se, low energy utilization requirement of low bandwidth (250 kbps) and low power (1 mW) radio]. The standard IPv6 header is 40 bytes (RFC 2460) and the entire 802.15.4 MTU is 127 bytes (IEEE), hence the actual data payload can become small, to be noted. To support interoperability applications need not know the constraints of the physical links through which their packets (datagrams) flow and can be achieved easily with fragmentation techniques. IP packets may be large compared to 802.15.4 max frame size, but IPv6 requires all links to support 1280-byte packets (RFC 2460), which is to be noted. Depending on the area of operation, 802.15.4 subnets may utilize multiple radio hops per IP hop, which is similar to LAN switching within the IP routing domain in Ethernet (Figure 11.2).

This standard supports extensive interoperability and other wireless embedded 802.15.4 network devices including on any other IP network link—Wi-Fi, Ethernet, GPRS, Serial lines, etc. It has already proven and established security—authentication, access control, and firewall mechanisms. In this way, network design, and policy determines access, as it has proven naming, addressing and translation, lookup, discovery, and proxy architecture for higher-level services—Network Address Translation (NAT), load balancing, caching, mobility, etc., may be planned easily. One can benefit from the well-established application-level data model and services like HTTP/HTML/XML, etc., application profiles and also the network management tools like Ping, Trace-route, SNMP, etc. In a single word, we need not re-invent many things in the wheel but can leverage the existing standards (Figures 11.3–11.5).

Octets				2	1	0/2	0/2/8	0/2	0/2/8	N	2
MAC Layer				Frame Control	Sequence Number	Destination PAN Identifier	Destination Address	Source PAN Identifier	Source Address	Pay Load Variable	Frame Check

Data Frame IEEE 802.15.4, MTU 127 Octets.

Octets	4	1	1	IEEE 802.15.4 MAC Layer - maximum 25 octets header + N octets pay load
PHY sub Layer	Preamble sequence	Start of frame Delimiter	Frame Length	MPDU
	SHR		PHR	<=============IEEE 802.15.4 MTU 127 Octets ==========================>
				PPDU

Figure 11.2 IEEE 802.15.4 PPDU detail.

Frame format uncompressed IPv6 / UDP, Worst Case Scenario

<===================================== MTU of 127 Octets===============================>

Preamble sequence	Start of frame Delimiter	Frame Length	MAC Header 23 octets or 44 octets	Dispatch Code	Uncompressed IPv6 Header of total 40 octets	Uncompressed UDP Header of 8 octets	Pay Load of 54 or 33 octets	FCS 2 octets
4 octets	1 octet	1 octet		1 octet				
<==PHY Sub Layer======>			<====MAC Sub Layer ====>		<==IP Layer (Network) ===> <Transport Layer=>			

1)	Dispatch code (01000001) indicates no compression
2)	Up to 54 / 33 octets left for payload with a max. size MAC header with null / AES-CCM-128 security

Figure 11.3 IEEE 802.15.4 IPv6 UDP frame worst-case detail.

To enable effective compression, LOWPAN_IPHC relies on information regarding the entire 6LoWPAN. LOWPAN_IPHC assumes that the following will be the common case for 6LoWPAN communication: Version is 6; Traffic Class and Flow Label are both zero; Payload Length can be inferred from lower layers from either the 6LoWPAN Fragmentation header or the IEEE 802.15.4 header; Hop Limit will be set to a well-known value by the source; addresses assigned to 6LoWPAN interfaces will be formed using the link-local prefix or a small set of routable prefixes assigned to the entire 6LoWPAN; addresses assigned to 6LoWPAN interfaces are formed with the Personal Area Network (PAN) derived directly from either the

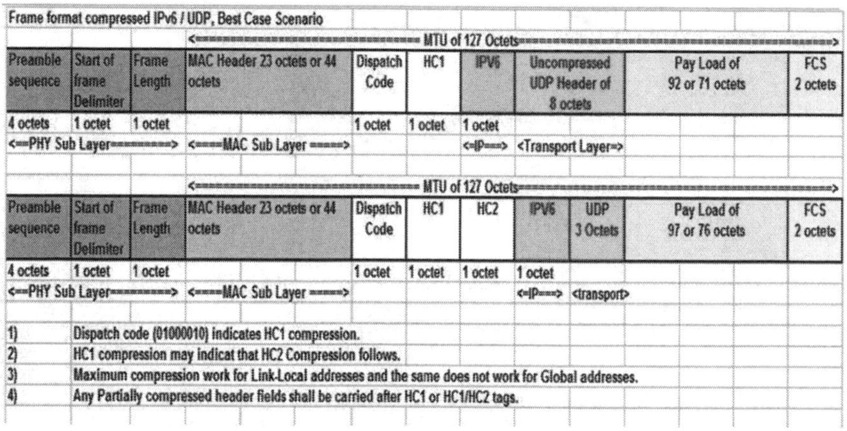

Frame format compressed IPv6 / UDP, Best Case Scenario									
			<============================ MTU of 127 Octets ========================>						
Preamble sequence	Start of frame Delimiter	Frame Length	MAC Header 23 octets or 44 octets	Dispatch Code	HC1	IPV6	Uncompressed UDP Header of 8 octets	Pay Load of 92 or 71 octets	FCS 2 octets
4 octets	1 octet	1 octet		1 octet	1 octet	1 octet			
<==PHY Sub Layer=========>			<=====MAC Sub Layer =====>			<=IP===>	<Transport Layer=>		

			<============================ MTU of 127 Octets ========================>						
Preamble sequence	Start of frame Delimiter	Frame Length	MAC Header 23 octets or 44 octets	Dispatch Code	HC1	HC2	IPV6	UDP 3 Octets	Pay Load of 97 or 76 octets	FCS 2 octets
4 octets	1 octet	1 octet		1 octet	1 octet	1 octet	1 octet			
<==PHY Sub Layer=========>			<=====MAC Sub Layer =====>				<=IP===>	<transport>		

1)	Dispatch code (01000010) indicates HC1 compression.
2)	HC1 compression may indicat that HC2 Compression follows.
3)	Maximum compression work for Link-Local addresses and the same does not work for Global addresses.
4)	Any Partially compressed header fields shall be carried after HC1 or HC1/HC2 tags.

Figure 11.4 IEEE 802.15.4 IPv6 UDP frame best-case detail.

6LoWPAN Dispatch Octet - Bit Pattern Description given below:

01 000001   uncompressed IPv6 addresses

01 000010   HC1 Compressed IPv6 header

01 010000   BC0 Broadcast header

01 111111   Additional Dispatch octet follows

10 xxxxxx   Mesh routing header

11 000xxx   Fragmentation header (first)

11 100xxx   Fragmentation header (subsequent)

Figure 11.5 IEEE 802.15.4 frame 6LoWPAN dispatch octet (control octet) bit pattern detail.

64-bit extended or the 16-bit short IEEE 802.15.4 addresses. To make it easy for the routing issues between "mesh" and "routing" topologies, there are two methods: "LOADng" standardized by the ITU under the recommendation ITU-T G.9903 and RPL standardized by the IETF ROLL working group.

Designers need to remember one thing before finalization of the IoT and 6LoWPAN frame/message structure: he or she is one of the many designers who are concurrently attempting the design. IoT device designers should not get into the tune of the Protocol or framework-related (ITU or other workgroups) designers, whose aim is to create and adhere to the international standards and meet all the boundary conditions. As IoT device designer is only planning to design a reliable and properly operating device within the limitations of the standards that are available, it is better to stick to one specific fame pattern to complete the solution. For example, one can finish his design even with an uncompressed IPv6 UDP worst-case frame model as shown above if the payload requirements fit into the 54 or 33 octets. Such an approach will help to recognize the design and solution uniquely in any global scenario and support easiness of design approach, focused frames, less complication, adherence to the international standards, etc.

11.2 6LoWPAN Node Roles and Routing

6LoWPAN nodes have special types and roles, such as nodes drawing their power from primary batteries, power-affluent nodes, mains-powered and high-performance gateways, data aggregators, etc. 6LoWPAN routing protocols should support multiple device types and roles.

6LoWPAN nodes are characterized by small memory sizes and low processing power, and they run on very limited power supplied by primary non-rechargeable batteries (a few KB of RAM, a few dozen KB of ROM/flash memory, and few MHz CPU are typical). A node's lifetime is usually defined by the lifetime of its battery.

Handling sleeping nodes is very critical in LoWPANs, more so than in traditional ad hoc networks. LoWPAN nodes might stay in sleep mode most of the time. Taking advantage of appropriate times for transmissions is important for efficient packet forwarding.

11.3 Routing Mechanism

Routing in 6LoWPANs possibly translates to a simpler problem than routing in higher-performance networks. LoWPANs might be either transit networks or stub networks. Under the assumption that LoWPANs are never transit networks (as implied by RFC4944), routing protocols may be drastically simplified. We will focus on the requirements for stub networks. Additional requirements may apply to transit networks depending on the model of the configuration.

The 6LoWPAN problem statement (RFC4919) briefly mentions four requirements for routing protocols:

- Low overhead on data packets
- Low routing overhead
- Minimal memory and computation requirements
- Support for sleeping nodes (consideration of battery savings)

Apart from a wide variety of conceivable routing algorithms for 6LoWPANs, it is possible to perform routing in the IP layer using a "route-over routing" approach or below IP, as defined by the 6LoWPAN format document (RFC4944) using the "mesh-under routing" approach.

The "route-over routing" approach relies on IP routing and therefore supports routing over possibly various types of interconnected links in general. Note: The "ROLL WG" workgroup is working for route-over approaches for Low-power and Lossy Networks (LLNs), not specifically for 6LoWPANs.

The "mesh-under routing" approach performs the multi-hop communication below the IP link. The most significant consequence of the mesh-under mechanism is that the characteristics of IEEE 802.15.4 directly affect the 6LoWPAN routing mechanisms, including the use of 64-bit (or 16-bit short) link-layer addresses instead of IP addresses. A 6LoWPAN would therefore be seen as a single IP link (Figure 11.6).

The designer's experience will be exhibited for the 6LoWPAN network— To avoid packet fragmentation and the overhead for reassembly, routing packets should fit into a single IEEE 802.15.4 physical frame, and application data should not be expanded to an extent that they no longer fit. It is very important to plan that the required result data should be fitting into a single frame of transmission.

Now let us examine the different stack properties of the 6LoWPAN. This routing requirements document states the routing requirements of

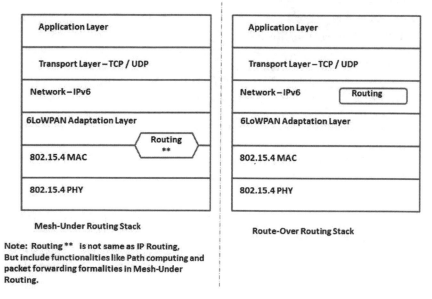

Mesh-Under Routing Stack

Note: Routing ** is not same as IP Routing,
But include functionalities like Path computing and
packet forwarding formalities in Mesh-Under
Routing.

Route-Over Routing Stack

Figure 11.6 "Mesh-under-routing" stack and "route-over-routing" stack.

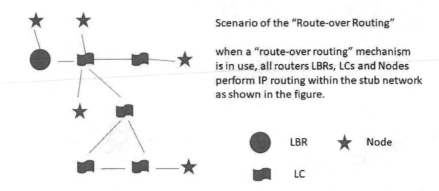

Scenario of the "Route-over Routing"

when a "route-over routing" mechanism
is in use, all routers LBRs, LCs and Nodes
perform IP routing within the stub network
as shown in the figure.

● LBR   ★ Node

▟ LC

Figure 11.7 Scenario of the "route-over routing."

6LoWPAN applications in general, providing examples for different cases of routing. It does not imply that a single routing solution will be favorable for all 6LoWPAN applications, and there is no requirement for different routing protocols to run simultaneously.

For multi-hop communication in 6LoWPANs, when a "route-over routing" mechanism is in use, all routers' LBRs, LCs, and nodes perform IP routing within the stub network as shown in Figure 11.7. In this case, the link-local scope covers the set of nodes within the symmetric radio range of a node.

When a LoWPAN follows the "mesh-under routing" configuration, the LBR is the only IPv6 router in the LoWPAN. This means that the IPv6 link-local scope includes all nodes in the LoWPAN. For this, a mesh-under mechanism MUST be provided to support multi-hop transmission (Figure 11.8).

It is to be noted that in both the above cases 6LoWPAN addresses are typically assigned based on the EUI-64 addresses assigned at manufacturing time to nodes, or based on a (from a topological point of view) more or less random process assigning 16-bit MAC addresses to individual nodes. Within a 6LoWPAN, there is no opportunity for aggregation or summarization of IPv6 addresses beyond the sharing of (one or more) common prefixes.

Not all devices that are within the radio range of each other need to be part of the same LoWPAN. When multiple LoWPANs are formed with globally unique IPv6 addresses in the 6LoWPANs, and device-A of LoWPAN-1 wants to communicate with device-B of LoWPAN-2, the normal IPv6 mechanisms will be employed. For route-over routing, the IPv6 address of device-B is set as the destination of the packets at LoWPAN-1, and the

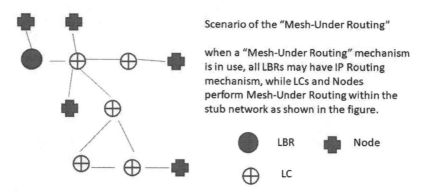

Scenario of the "Mesh-Under Routing"

when a "Mesh-Under Routing" mechanism is in use, all LBRs may have IP Routing mechanism, while LCs and Nodes perform Mesh-Under Routing within the stub network as shown in the figure.

Figure 11.8 "Mesh-under-routing" scenario.

devices perform IP routing to the LBR for these outgoing packets. For mesh-under routing, there is one IP hop from device-A to the LBR of 6LoWPAN-1, no matter how many radio-hops they are apart from each other. This, of course, assumes the existence of a mesh-under routing protocol to reach the LBR. Note that a default route to the LBR could be inserted into the 6LoWPAN routing system for both route-over and mesh-under approaches.

Chapter 12

IoT Design Components

One of the biggest challenges facing today's and tomorrow's product manufacturers is the demands of the IoT on their business. As the cloud economy booms, software development continues to thrive, and electronics get smaller and smaller, we continue toward a future where everything from your toothbrush to your car tire is becoming connected. This new reality is forcing traditional product companies to go beyond and create a website or online customer service portal for their customers. This new change is demanding brands of all sizes, to understand more about the three things that enable IoT products: hardware, software, and connectivity.

12.1 Hardware

Let us start with hardware. There are hundreds of hardware companies that are very much attracted to the IoT explosion. Many companies offer a broad portfolio of microcontrollers, processors, sensors, and wireless connectivity technologies that allow business organizations to develop IoT applications specific to their unique requirements. Let us primarily examine a few companies.

Texas Instruments (TI): With a market of about US$55 billion, TI offers processors and microcontrollers, connectivity technologies, sensors, analog, and digital ICs. The important target business verticals are Wearables, Manufacturing, Building Automation Healthcare Smart Cities, and Automotive. TI IoT building blocks include front-end nodes such as power management, analog signal chain and advanced sensing technologies generating precise data, hybrid gateways or bridges for IoT data transmission, and a vast cloud infrastructure offering third-party IoT cloud services.

TI Products: Supported processor families: RM48, TMS570, ARM Cortex-M4F MSP432, MSP430, MSP430X, Stellaris (ARM Cortex-M3, ARM Cortex-M4F), etc.

TI Supported Tools: Rowley CrossWorks, IAR, GCC, Code Composer Studio

Intel: With a market of about US$80 billion, Intel offers end-to-end IoT solutions connecting smart sensors and devices across the cloud infrastructure and instigating automated machine operations. The Intel IoT platform provides a foundation for third-party applications to connect and perform tailored IoT functionality in a secure environment specific to unique customer requirements. Intel processing chips capture sensor data and verify and transmit raw information to the cloud database yielding actionable insights using advanced Intel analytics capabilities.

Intel Products: Intel® architecture scales for the IoT through a wide range of product offerings. Intel® Quark™, Intel Atom™, Intel Core™, and Intel Xeon® processors each support a wide range of performance points with a common set of code. Analytics, encryption, and new application requirements in IoT are driving the need for high levels of performance and computing headroom. For every case, there's an Intel processor.

Note that Intel is offering end-to-end IoT solutions which include all the necessary hardware, software, and connectivity-related components through Intel® Internet of Things Solutions Alliance.

Qualcomm: With a market of about US$150 billion, as a global leader in mobile processor and connectivity technologies, Qualcomm delivers off-the-shelf IoT products for a range of industry verticals. The company is inclined toward consumer-oriented IoT applications with initiatives such as smart homes and smart cities. Qualcomm IoT products allow consumers to transform the existing home network and power infrastructure into a connectivity hub for mobility-based IoT use-cases. Qualcomm products for automotive, mobile, healthcare, and education allow organizations from these verticals to develop consumer-facing IoT applications. Notable products include the Snapdragon processor powering over a billion mobile devices globally, wireless connectivity systems, and advanced software IoT applications that allow business organizations to empower their customers with next-generation IoT capabilities.

Qualcomm Products: Qualcomm MDM9207-1, Qualcomm Snapdragon 650, etc., integrated with ARM Cortex A7 AP and GNSS—for location services and related products.

ARM: With a market of about US$24 billion, ARM operates on a distinct business model of creating and licensing its technologies as intellectual property to third-party IoT vendors. Licensees of ARM processors for IoT applications range from consumer technology giants Apple and Samsung to enterprise technology vendors IBM, Qualcomm, and Intel, among others. ARM IoT offerings range from sensors and processing chips to networking equipment and cloud servers establishing an end-to-end IoT hardware infrastructure. The company also provides development platforms such as the ARM mbed solution for third-party IoT vendors and business organizations to create IoT applications based on ARM microcontrollers.

ARM Products: ARM mbed IoT Device Platform, ARM Cortex-M based CPU, etc. ARM mbed IoT device platform provides all the key ingredients to build secure and efficient IoT applications through ARM's mbed OS, mbed Device Server, and mbed Community Ecosystem.

There are other important companies like **Atmel** with about US$3.1 billion, **Freescale** with about US$4.2 billion, which are supporting the IoT solutions embedded processors, microcontrollers, and multi-faceted sensors used in industrial, healthcare, automotive, smart energy as well as consumer applications.

Besides, **Actel (now Microsemi), Altera, Cortus, Infineon, Microchip, NEC, NXP, Silicon Labs, Xilinx, etc.,** are a few notable names in the IoT hardware business. The designers need to understand the importance of the hardware they identify for the targeted IoT design.

Differentiation has to be made whether they are making a 6LoWPAN node or 6LoWPAN Local Controller (LC) or 6LoWPAN Border Router and accordingly the hardware and the future expansion and growth path are to be estimated by the designers.

Note: Always designers should plan for long-term stability, survival, and reusability aspects of hardware like Processor IC, Radio, etc., for an IoT device. Hardware must be reusable, and different software may have to be re-deployable for different market scenarios and requirements. Hence, identifying a good and durable processor for all the hardware targeted shall exhibit the designer's capabilities—one of the important marketing and business aspects.

12.2 Software

As part of the software, we need to look into the available and suitable operating systems (OS) for IoT design, and other related stacks and software platforms or frameworks for the design considerations, as we have already discussed Middleware etc. We would discuss only few OS that in a way are more popular in the market, and useful for target IoT implementation (refer Figure 12.1).

FreeRTOS is designed to be small and simple. The kernel itself consists of only three or four C files. It provides methods for multiple threads or tasks, mutexes, semaphores, and software timers. Key features include a very small memory footprint, low overhead, and very fast execution. IoT-LAB uses FreeRTOS by default for basic development for WSN430 and M3 nodes.

Contiki is an open-source OS for networked, memory-constrained systems with a particular focus on low-power wireless IoT devices. Contiki provides three network mechanisms: the uIP TCP/IP stack, which provides IPv4 networking; the uIPv6 stack, which provides IPv6 networking; and

CPU --->	WsnWSN430 Node	M3 Node	A8 Node
OS			
FreeRTOS	X	X	-
Contiki	X	X	-
Riot	X	X	-
TinyOS	X	-	-
OpenWSN	X	X	-
Embedded Linux	-	-	X

Figure 12.1 Operating systems useful for IoT device design.

the Rime stack, which is a set of custom lightweight networking protocols designed specifically for low-power wireless networks. The IPv6 stack was contributed by Cisco and was, at the time of release, the smallest IPv6 stack to receive the IPv6 Ready certification. To run efficiently on memory-constrained systems, the Contiki programming model is based on protothreads. A protothread is a memory-efficient programming abstraction that shares features of both multi-threading and event-driven programming to attain a low memory overhead of each protothread.

RIOT is a real-time multi-threading OS that explicitly considers devices with minimal resources but eases development across the wide range of devices that are typically found in the IoT. RIOT is based on design objectives including energy efficiency, reliability, real-time capabilities, small memory footprint, modularity, and uniform API access, independent of the underlying hardware (this API offers partial POSIX compliance). Several libraries (e.g., Wiselib) are already available on RIOT, as well as a full IPv6 network protocol stack including the latest standards of the Internet Engineering Task Force (IETF) for connecting constrained systems to the Internet (6LoWPAN, IPv6, RPL, TCP, and UDP).

TinyOS is a component-based OS and platform targeting wireless sensor networks. TinyOS is an embedded OS written in the nesC programming language as a set of cooperating tasks and processes. TinyOS programs are built out of software components, some of which present hardware abstractions. Components are connected using interfaces. TinyOS provides interfaces and components for common abstractions such as packet communication, routing, sensing, actuation, and storage.

OpenWSN project is an open-source implementation of a fully standards-based protocol stack for capillary networks, rooted in the new IEEE802.15.4e Time-Slotted Channel Hopping (TSCH) standard. IEEE802.15.4e, coupled with IoT standards, such as 6LoWPAN, RPL, and CoAP, enables ultra-low power and highly reliable mesh networks which are fully integrated into the Internet. The resulting protocol stack will be a cornerstone to the Internet of (Important) Things.

Embedded Linux is created using OpenEmbedded, the build framework for embedded Linux. OpenEmbedded offers a best-in-class cross-compile environment. Only A8 nodes are powerful enough to support an embedded Linux.

As we had a detailed discussion on middleware (Chapter 9) IoT networks are made of heterogeneous devices to collect different specific inputs regularly. Hence, the supporting middleware should shield their heterogeneity and seamlessly communicate with Services and Application levels by the way of delivering different required information packets which shall enable the security and scalability of IoT architecture. It is to be noted that some of the protocol stacks are in-built with required middleware extensions and they are accompanying the OS itself.

Chapter 13

IoT Design Implementation

With the hope that we could understand some basics about IoT devices, let us attempt the design of a 6LoWPAN LC as well the 6LoWPAN Node. For the sake of design completion, we shall talk about 6LoWPAN Border Router (LBR). LBR is full function router that demarks the LoWPAN network from the external Internet world. We shall not be concentrating and working on the LBR presently.

Hardware is the heart of the design consideration and we have already identified many companies and different microcontrollers. To start with as an example, one consider the Texas Instruments (TI) CC-6LoWPAN-DK-868 development kit, which includes CC1180 Network Processor and CC430 SoC, etc. The details of the kit include the following: 2 Nos of CC1180DB nodes (CC1180 NWP plus MSP4305438A host MCU); 2 Nos EM430F5137900 (CC430 SoC) nodes; 1 OMAP-L138-based Edge Router Board (Gateway, running Linux); 1 Adapter board, for connection of CC1180EM to OMAP-L138 board; 1 CC1180EM (radio interface to OMAP-L138 board); MSP-FET430UIF Debugger, used to debug and download code to nodes; Ethernet Cable (Crossover, for direct connection to PC); RS-232 NULL modem cable (used for Linux debug console); Power supply for OMAP-L138 board, including cables; USB cable for MSP-FET430UIF Debugger; Antennas, for 868/915 MHz band; Batteries (including holders for CC430 boards); Quick Start Guide; Sensinode Node View 2.0 Network Analyzer PC SW; software application examples.

At this juncture, it is to be noted that if one wishes to attempt the LBR (LoWPAN Border Gateway), the hardware considerations may be different as it needs A8 kind of MCU capabilities, and Embedded Linux kind of OS.

NOTE: This does not mean that one needs to use different MPU devices for each of the LC, RFD, FFD, LBR, etc., as a must; and that is where the designer's experiences will be exhibited—based on the experience and targeted design considerations the designer needs to choose such MPU device that may fit into all these different requirements—shall help the design to survive for long-term multifaceted strategies. In present-day, many different software and OS options that may suit a single MPU device are part of the designer's choice.

FreeRTOS and Nanostack shall work on most of these MCUs. Hence, we assume that the 6LoWPAN Node (or simply Node) can be implemented on WSN430 MCU or equivalent, while 6LoWPAN LC shall be implemented on M3 MCU or equivalent (ARM7 Cortex M3). Now let us look into the more functional understanding of the design implementation and the possible different OS, etc.

13.1 Using FreeRTOS and Nanostack

Currently, there are few known open-source implementations of 6LoWPAN stacks. Based on the survey report we have chosen one of the open-source 6LoWPAN implementations called "Nanostack." Among the available open-source 6LoWPAN implementations, we have selected Nanostack due to two strong facts. First, the stack operates on top of a real-time kernel FreeRTOS, making it suitable for time-critical tasks like real-time control or network synchronization. Second, the stack implements mesh-under routing in contrast to other stacks that adopt route-over routing.

Nanostack (v1.1) is an implemented and working 6LoWPAN stack distributed under a General Public License (GPL) license. In this, we focus on the open-source Nanostack release, version 1.1, and hereafter refer to it simply as Nanostack. Nanostack implements most of the RFC4944, namely, processing of Mesh-Header, Dispatch-Header (and its alternatives LOWPAN_HC1, LOWPAN_BC0, and IPv6), mesh-under routing and compression of IP and UDP headers.

Nanostack operates on top of the FreeRTOS kernel, (refer to Figure 13.1), a multi-platform, mini, real-time kernel, which in turn runs in a microcontroller located in a hardware platform. In the context of WSN, the hardware platform usually incorporates sensors and wireless radio circuitry and receives the name of "wireless sensor node" or "sensor node" among other names used in the literature.

In Nanostack each layer is associated with a module of code. Nanostack modules can be enabled or disabled to modify the functionality of the stack. For example, we describe an optional module called Network Router Protocol (NRP) that can be enabled to make a node work as Gateway. To simplify the description of the stack, the NRP module is not considered until Gateway (Figure 13.1). Nanostack contains also an Internet Control Message Protocol (ICMP) module that introduces ICMP layer functionality. ICMP is a protocol used for communicating exceptional messages between nodes or layers in the stack; ICMP logic can be and will be distributed in several layers, and its inclusion can compromise the simplicity and structure that we intend to adopt in the description of this complex stack, that is, one can include or exclude the ICMP depending on the interest. This

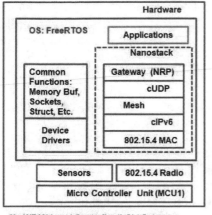

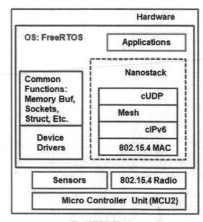

6LoWPAN Local Controller (LC) / Gateway   **6LoWPAN Node**

Note: Please observe the hardware difference MCU1 and MCU2 in the above, and also the network Stack layers.

Figure 13.1 FreeRTOS Nanostack architecture.

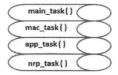

Figure 13.2 Few Nanostack functions.

combination (FreeRTOS and Nanostack) implementation currently does not support 6LoWPAN fragmentation and, to some extent, unicast and multicast address mapping is supported.

FreeRTOS is a multitasking OS. Nanostack was designed to make use of this feature and it executes tasks dedicated to different purposes concurrently. Typical Nanostack tasks running in a sensor node are shown (Figure 13.2). The system task labeled **stack_main**() is the core task of Nanostack; the task labeled **mac_task**() processes packets at the MAC layer level. Task **app_task**() is a task dedicated to executing the user's code at the application layer level. Finally, task **nrp_task**() is a task that is executed when the NRP module for Gateway functionality is enabled.

Also note that the application layer is positioned outside and is independent. It is independent of the Nanostack in the sense that it is not related to the dispatcher and instead receives/sends packets by calling socket functions. The application layer consists of an independent FreeRTOS task that implements the socket API to access the stack. Incoming packets are redirected to the queue *socket* inside the socket structure and are then retrieved at the application layer by calling the **socket_read**() function. Outgoing packets are generated at the application layer by invoking function **socket_sendto**(), which sets **buf. dir** = BUFFER_OUT in the instance *b* that holds the packet, and finally are pushed to queue *events* by calling **stack_ buffer_push**().

Packets move through the stack using instances of structure **mem_buffer_t**. (refer Figure 13.3). The important observation one has to do is that even when we say that there is a movement of packets inside the stack, Nanostack actually never moves packets but merely pointers are changed to instances of **mem_***buffer_t* which is part of a ring buffer implementation.

Incoming packets are directly stored in **mem_***b.buf*. Subsequent processing in the different layers of the stack parses the packet and fills accordingly the fields of **mem_buffer_t** structure with the information extracted from the packet. Outgoing packets are constructed layer by layer – each layer adding an appropriate header in **mem_b.buf**. In the stack, packets are handled employing instances of the data structure **mem_***buffer_t*. **mem_** *buffer_t* is defined in file **mem_***buffer.h* and reproduced below.

We indicate **buf** an instance of **mem_buffer_t**. The content of a packet is stored in the **buf** field of **mem_buffer_t**: **buf. buf**, where the notation expresses a qualified name (in this case, **buf** is an instance of **mem_buffer_t** and **buf** a field belonging to **buf**). In the actual implementation, **buf** is a pointer to an instance of **mem_buffer_t**, and a C language notation **buf–>buf** is more appropriate, although not used here to simplify the

```
Typedef struct {

        Struct socket_t*      socket;
        Socketaddr_t          dst_add;
        Socketaddr_t          src_add;
        Module_ID_t           from;
        Module_ID_t           to;
        Buffer_dir_t          dir;
        unit16_t              buf_ptr;
        unit16_t              buf_end;
        unit16_t              size;
        buffer_options_t      options;
        unit8_t               buf(2);

} mem_buffer_t;
```

Figure 13.3 Memory buffer structure.

Note: In the above structure we have shown only the required fields of a buffer instance or structure, but any designer may add one or two more fields of his requirement, based on his design choice. These structures will give confidence to the designers that they can start with, and in the course, one can add more fields if they feel so.

notation. Linux kernel connoisseurs can imagine **mem_buffer_t** as a structure analogous to the socket buffer structure **sk_buff** in the Linux kernel.

Three fields in **mem_buffer_t** of particular importance are **buf. from, buf. to**, and **buf. dir**. Fields **buf. from** and **buf. to** indicate the previous and posterior layers that processed and will process the packet, respectively. Field **buf. dir** indicates the direction of movement of the packet; the value **BUFFER_IN** indicates an incoming packet moving toward the Application Layer whereas the value **BUFFER_OUT** indicates an outgoing packet moving toward the Network Layer.

The pool of pre-allocated instances (refer Figure 13.4), is managed to utilize a ring buffer labeled **stack_buffer_pool[]**. **stack_buffer_pool[]** is indexed by two position pointers: a pointer indicating a reading position (**stack_buffer_RD**) and a pointer indicating a writing position (**stack_buffer_WR**), indicating the next available place where a **mem_buffer_t** instance can be taken or returned, respectively.

The functions **stack_buffer_get()** and **stack_buffer_free()** are front-en d functions that will take or return a **mem_buffer_t** instance from the pool and at the same time these functions initialize or reset the fields of the structure. Behind these functions, **stack_buffer_pull()** and **stack_buffer_add()** perform the actual unload/load of **mem_buffer_t** pointers from/to the pool of instances **stack_buffer_pool[]**.

In the above implementation of the stack, a call to **stack_buffer_get()** is always followed by a call to **stack_buffer_free()** at some point in the life of the packet to recycle the finite number of instances **mem_buffer_t**.

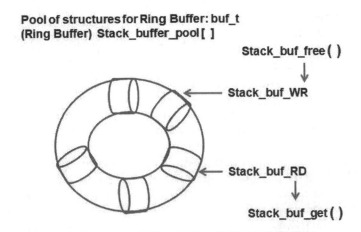

Pool of structures for Ring Buffer: buf_t (Ring Buffer) Stack_buffer_pool[ ]

Stack_buf_free ( )

Stack_buf_WR

Stack_buf_RD

Stack_buf_get ( )

Buffer Mechanism for Ring buffer buffer_t Instances

Figure 13.4 Ring buffer mechanism.

NOTE: Before we work deep into the stack layers, we should note that in the above we used FreeRTOS and Nanostack-based implementation. We can use other OS and stacks for the same implementation also. For example, in the below, we will use the implementation using the Contiki OS and Contiki 6LoWPAN stack. Designers can choose their choice and known OS and the stack that are useful for easy implementation. Hence, the implementation from one designer to other may defer, but the target goal "design of IoT" is always met. Also note that "Ring buffer" implementation is one way of memory saving method and can be used with both OS (FreeRTOS and Contiki) cases as well as with any other OS.

13.2 Using Contiki OS and Contiki 6LoWPAN Stack

Contiki is an open-source OS for memory-constrained embedded systems and wireless sensor networks. It is highly portable and ported to more than 12 different microprocessor and microcontroller architectures. Contiki is designed for microcontrollers with very limited memory size. A typical Contiki configuration requires 2 KB of RAM and 40 KB of ROM.

Contiki provides IP communication, both for IPv4 and IPv6, through uIP and uIPv6 stacks. Its IP stack allows it to communicate directly with other IP-based applications and web services, including Internet-based services. Contiki supports many other protocols and mechanisms like 6LoWPAN header compression, RPL routing, and CoAP application layer protocol. It has also a low-power radio networking stack called **Rime** that can be used for sensor network communication. The Rime protocol stack provides a set of communication primitives, ranging from best-effort local neighbor broadcast and reliable local neighbor unicast to best-effort network flooding and hop-by-hop reliable multi-hop unicast. Applications or protocols running on top of the Rime stack may use one or more of the communication primitives provided by this stack. Contiki provides three simulation environments: the MSPsim emulator, the Cooja cross-layer network simulator, and the Netsim process-level simulator. They can help in the development and debugging of software, before testing it on the hardware.

Contiki supports two programming models: a "multi-threaded programming" model and an "event-driven programming" model. In multi-threaded programming, each program has its specific thread. Different threads can run alongside each other in the system and thus share the same microprocessor. The OS coordinates the execution of the threads to give each thread time to run on the microprocessor. The drawback of multi-threaded programming is that each thread needs its respective piece of memory to hold the state of the thread, the so-called stack of the thread. The thread keeps in its stack the set of local variables and returns values of the functions it called. The problem is that the amount of stack memory a thread needs is unknown in advance, so the allocated stack memory is typically oversized with a comparatively large amount of unused memory.

With the event-driven programming of Contiki, the program is composed of event handlers. An event handler is a short section of code that describes how the system responds to events. When an event occurs, the system executes the corresponding event handler. As part of event-driven programming, Contiki supports protothreads and they can be seen as a combination of events and threads. From threads, protothreads have inherited the blocking wait semantics. From events, protothreads have inherited the "stacklessness" and the low memory overhead. The blocking wait semantics allow linear sequencing of statements in event-driven programs. In Contiki, processes can be implemented as protothreads running on top of the event-driven Contiki kernel. When a process receives an event, the corresponding protothread is invoked. The event may be a message from another process, a timer event, a notification of sensor input, or any other type of event in the system. Processes may wait for incoming events using the protothread conditional blocking statements.

Figure 13.5 shows the typical differences between (a) FreeRTOS and Nanostack and (b) Contiki OS and Contiki 6LoWPAN stack. Note that the major change is with Contiki that one can support the fragmentation of packets, which shall become a must for complete full IPv6 stack implementation. However, all may not need the full implementation of the IPv6 and if they can live with the route-under mesh, then one can go for FreeRTOS and Nanostack, as per the designer's choice.

Hardware & Software (OS & Stack) 6LoWPAN Implementation Options					
	Gateway	Node	Gateway	Node	
	Application	Application	Application	Application	
	BGP / ... / ...	---------	NRP	---------	
Contiki OS	TCP	TCP	UDP	UDP	FreeRTOS
	IP CCN	IP CCN	Mesh	Mesh	
Common functions Memory, Buffers, Clock, etc.	CSMA	CSMA	-NIL- **	-NIL- **	Common functions Memory, Buffers, Clock, etc.
	Contiki MAC	Contiki MAC	6LoWPAN MAC	6LoWPAN MAC	
Device Drivers	6LoWPAN PHY	6LoWPAN PHY	6LoWPAN PHY	6LoWPAN PHY	Device Drivers
Contiki OS & Contiki 6LoWPAN stack. Fragmentation of Packets Possible. Route-Under Mesh and Rout-Under Route Both Possible. Content Centric Networking (CCN)			FreeRTOS & Nanostack. ** Fragmentation of Packets NOT Possible. Route-Under Mesh only Possible		
Hardware / MCU / 802.15.4 Radio /					

Figure 13.5 6LoWPAN—HW, OS, SW, etc., implementation options.

Chapter 14

Design of Layered Architecture

Now let us deep dive into different layers of the stack. Depending on the layer we may jump between the two different OS (Contiki and FreeRTOS) implementations now and then (Figure 14.1), that enhance the designer's understanding, and the designer has to use only one OS and stack—either Contiki or FreeRTOS—depending on his choice. In the below, as we proceed from layer to layer, we shall be highlighting the important events and actions that are to be observed by designers to implement.

14.1 Flow/Route of the Message

An incoming packet enters the stack in the MAC layer through a Function Call (refer Figure 14.2). This is facilitated by the respective device driver functions. In this layer (MAC), a data packet is stored inside a Buffer structure. Additional values are stored in required buffer fields, for example, BUFFER_IN to indicate that the packet will move toward the application layer in the stack. Then the instance Buffer Pointer is pushed to queue *events* by calling another function *per se* **Stack Push** which in turn calls **Queue Send** to input the instance into the queue.

The main idea, if implemented similarly, is that an instance buffer pointer travels among the required layers, and during its journey the sub-fields of Buffer acquire the information stored in the packet by decomposing it (**incoming case: BUFFER_IN**) or, conversely, the layers utilize the information in the fields of Buffer to compose the packet (**outgoing case: BUFFER_OUT**). The movement between layers is facilitated by every time pushing the instance to **queue** *events* and letting the dispatcher deliver the instance to the appropriate layer by doing a function call to a dedicated handler function in the layer, and passing the instance as a parameter.

Incoming packet leaves the MAC layer and enters **queue events**; it is delivered by the dispatcher (or one can call as the main program) to the next higher layer in the **BUFFER_IN** direction. In the present explanation the next upper layer is IP layer (**Nanostack—CIP or MESH; Contiki—uIP or**

DOI: 10.1201/9781003303206-14

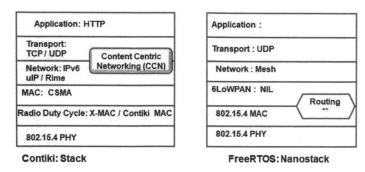

Figure 14.1 Contiki and FreeRTOS stack layers.

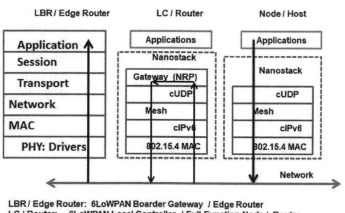

Figure 14.2 6LoWPAN message flow.

Rime). The layers in the stack are modular in the sense that at compilation time some layers can be added or removed from the stack.

IP layer will decide the next destination layer in the stack. An incoming packet targeted for the current node will move in **BUFFER_IN** direction towards the application layer, traveling through the Transport layer (**Nanostack—cUDP; Contiki—TCP/UDP**). When a packet is targeted to another node, it may move towards the lower layer **BUFFER_OUT** direction to obtain routing information, and from there continue **BUFFER_OUT** to the MAC layer for forwarding. For outgoing packets, the transition involves the displacement of Buffer in the **BUFFER_OUT** direction through the layers transport, IP, and MAC, and then device drivers to place the packet onto the network.

In essence, the processing of the packet in each layer will determine its final fate according to different situations and contexts. The different situations and decisions mandated by the logic in the specific stack and the designer's choice of implementation shall become the final target. In Figure 14.2, at the LC the different destination packets will not reach the application layer, they may be sent to LBR, from the NPR layer itself, which is one of the examples of the implementations.

14.2 Contiki CCN Layer

Contiki layered architecture and design implementation are based on the **Content-Centric Networking (CCN or CCNx)** communication layer based on named data. CCNx protocol is a transport protocol for the CCN communication architecture. According to the CCN specifications, it is built on named data where the content name replaces the location address. The CCNx protocol provides location-independent delivery services for named data packets. The services include multi-hop forwarding for end-to-end delivery, flow control, transparent and automatic multicast delivery using buffer storage available in the network, loop-free multipath forwarding, verification of content's integrity regardless of delivery path, and carriage of arbitrary application data. Although the CCNx protocol is designed to deliver content based on their names, applications can run it on top of UDP or TCP to take advantage of existing IP connectivity. Since content is named independently of location in the CCNx protocol, it may also be preserved indefinitely in the network. Every packet of data may be cached at any CCNx router. Providing support for multicast or broadcast delivery, the network's use is more efficient when many people are interested in the same content. In general, CCN is specific to Contiki OS only.

A CCN name is a hierarchical name attributed to content. It simply contains a sequence of components of arbitrary lengths. There are no restrictions on what byte sequences may be used. The implemented communication layer specifies only the name structure and does not assign any meanings to names. It is up to applications or global naming conventions to set and interpret meanings given to names. Application developers are free to design their custom naming conventions.

In the example given below (refer Figure 14.3), the prefix/Temperature/Amaravati/CM-Office/identifies the set of contents indicating the temperatures in the different floors and offices of Chief Minister's Offices located at the Amaravati Capital City of Andhra Pradesh.

A CCN name is a hierarchical name attributed to content. It simply contains a sequence of components of arbitrary lengths. There are no restrictions on what byte sequences may be used. The implemented communication layer specifies only the name structure and does not assign any

Prefix: /Temparature/Amaravati/CM-Office

Names of Contents:

/Temparature/Amaravati/CM-Office/Floor1/	Office 115	:23 deg. C
/Temparature/Amaravati/CM-Office/Floor1/	Office 116	:22 deg. C
/Temparature/Amaravati/CM-Office/Floor1/	Office 117	:24 deg. C
/Temparature/Amaravati/CM-Office/Floor1/	Office 118	:23 deg. C
/Temparature/Amaravati/CM-Office/Floor2/	Office 215	:21 deg. C
/Temparature/Amaravati/CM-Office/Floor2/	Office 216	:22 deg. C
/Temparature/Amaravati/CM-Office/Floor2/	Office 217	:22 deg. C

Figure 14.3 Contiki CCNx naming conventions.

meanings to names. It is up to applications or global naming conventions to set and interpret meanings given to names. Application developers are free to design their custom naming conventions. To represent names, we refer to the Uniform Resource Identifier (URI) scheme (RFC 3986: URI Generic Syntax). A URI is a sequence of characters identifying a physical or abstract resource. The URI syntax is organized hierarchically, with components listed in the order of decreasing significance from left to right. A URI is composed of a limited set of characters consisting of digits, letters, and a few graphic symbols. A reserved subset of those characters may be used to delimit syntax components within a URI, while the remaining characters, including both the unreserved set and those reserved characters not acting as delimiters, define each component's identifying data. When producing a URI from a CCN name, only the generic URI unreserved characters are left unescaped. These are the US-ASCII upper- and lowercase letters (A – Z, a -z), digits (0–9), and the four special characters that are "period" (.) under-score (_), tilde (~), and hyphen (-). All other characters are escaped using the percent-encoding method of the URI Generic Syntax.

At this juncture for clear understanding, we wish to remind you of 6LoWPAN frame-related information which has been discussed earlier. LOWPAN_IPHC relies on information about the entire 6LoWPAN. LOWPAN_IPHC assumes that the following will be the common case for 6LoWPAN communication: Version is 6; Traffic Class and Flow Label are both zero; Payload Length can be inferred from lower layers from either the 6LoWPAN Fragmentation header or the IEEE 802.15.4 header; Hop Limit will be set to a well-known value by the source; Addresses assigned to 6LoWPAN interfaces will be formed using the link-local prefix or a small set of routable prefixes assigned to the entire 6LoWPAN; addresses assigned to 6LoWPAN interfaces are formed with an Interface Identifier (IID [RFC 4291], 64-bit prefix part of, IPv6 128-bit address; or 16-bit short address) derived directly from either the 64-bit extended or the 16-bit short IEEE 802.15.4 addresses.

6LoWPAN Dispatch Octet—Control Bit Pattern Description given below:

01 000001 uncompressed IPv6 addresses
01 000010 HC1 Compressed IPv6 header

01 010000 BC0 Broadcast header
01 111111 Additional Dispatch octet follows
10 xxxxxx Mesh routing header
11 000xxx Fragmentation header (first)
11 100xxx Fragmentation header (subsequent)

14.3 6LoWPAN MAC Layer

Contiki MAC is designed to be simple to understand and implement. Contiki MAC uses only asynchronous mechanisms, no signaling messages, and no additional packet headers. Contiki MAC packets are ordinary link-layer messages. Contiki MAC has a significantly more power-efficient wake-up mechanism than previous duty cycling mechanisms. This is achieved by precise timing through a set of timing constraints. Also, Contiki MAC uses a fast sleep optimization, to allow receivers to quickly detect false-positive wake-ups, and a transmission phase-lock optimization, to allow run-time optimization of the energy efficiency of transmissions. Contiki MAC is a radio-duty cycling protocol that uses periodical wake-ups to listen for packet transmissions from neighbors. If a packet transmission is detected during a wake-up, the receiver is kept on to be able to receive the packet. When the packet is successfully received, the receiver sends a link-layer acknowledgment. To transmit a packet, a sender repeatedly sends its packet until it receives a link-layer acknowledgment from the receiver. Packets that are sent in broadcast mode do not result in link layer acknowledgments. Instead, the sender repeatedly sends the packet during the full wake-up interval to ensure that all neighbors have received it. The principal mechanism of Contiki MAC is shown below (Figure 14.4).

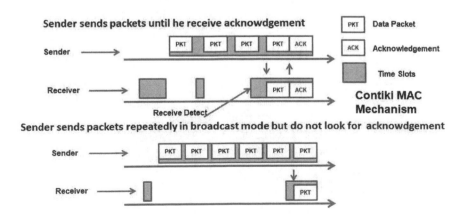

Figure 14.4 Contiki MAC message flow mechanism.

For **Nanostack MAC (FreeRTOS)** the processing of a packet at the MAC layer level comprises the identification and incorporation of the MAC headers in the MAC sub-layer part of an IEEE 802.15.4 data frame and the logic related to retransmissions when transmission acknowledgments (ACKs) are expected. The actual portion of data that we call *packet* is the *MAC protocol data unit* (**MPDU**) (**PSDU at PHY layer level**) of the IEEE 802.15.4 Data Frame. This portion of data is the one, pointed by the **buf. buf** field of the **mem_buffer_t** instance. The logic in the MAC layer is distributed in two different parts of the code, in the MAC layer module from the stack, and in a dedicated task labeled **mac_task()** that is initialized with the stack. The incoming packet enters the stack in the MAC layer through a call to **mac_push()**. In this layer, a data packet is stored inside a **mem_ buffer_t** structure, in the field **buf. buf**. Additional values are stored in **buf** fields, for example, **buf. dir = BUFFER_IN** to indicate that the packet will move towards the application layer in the stack. Then the instance **buf** is pushed to queue *events* by calling function **stack_buffer_push()**, which in turn calls **Queue_Send()** to input the instance into the queue.

Figure 14.5 depicts the MAC layer module from the stack. It can be observed in the figure that there is no processing of packets in the **BUFFER_ IN** direction (**toward the application layer**). The processing in this direction is instead carried out in the dedicated task **mac_task()** that among other

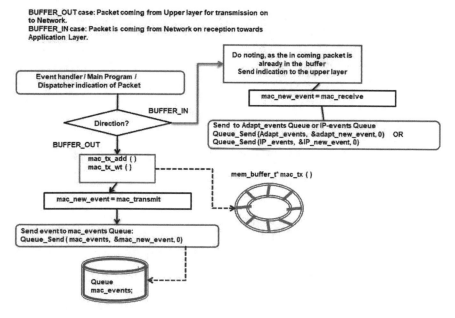

Figure 14.5 MAC Layer message flow.

jobs takes care of processing incoming packets. In the below, we discuss the processing of a packet in the **BUFFER_IN** and **BUFFER_OUT** (**toward network**) directions considering both, the MAC layer module from the stack and the dedicated task **mac_task**() running concurrently in the stack.

Note that in the above MAC layer actions, only the pointer of the buffer shall be changed from one queue to another queue once the **mac_tx_add**() function does the necessary changes in the buffer using the **mac_tx_wt**() function. In the above case, the buffer is attached to the "**Queue_Send**." This shall save a lot of memory space and unnecessary creation of buffer space again and again.

14.4 6LoWPAN Adaptation Layer

This layer may not be present in FreeRTOS and Nanostack implementations, while Contiki-based implementation can support this adaptation layer. Primary actions of this layer are "Fragmentation and Reassembly," "Header Compression," and "Routing."

Fragmentation and Reassembly: Maximum transmission unit (MTU) for IEEE 802.15.4 is 127 bytes. A frame can have 25 bytes, a header, a footer, and addressing overheads. Additionally, the security header imposed by the Link layer adds 21 bytes overhead when AES-CCM-128 is used. As a result, the remaining payload is 81 bytes. As we all know IPv6 packet size is a minimum of 1280 bytes and that this size is larger than IEEE 802.15.4 frame. In this condition, IPv6 packet size is unable to be encapsulated in one IEEE 802.15.4 frame, and hence Fragmentation and Reassembly become the need of the hour. IPv6 frame that needs to be transmitted over IEEE 802.15.4 frame has to be divided into more than 16 fragments. Hence, adaptation layer should handle such fragmentation and reassembly process.

Header Compression: IEEE 802.15.4 defines four types of frames: beacon frames, MAC command frames, acknowledgment frames, and data frames. IPv6 packets must be carried on data frames. After the packet is fragmented and transmitted over IEEE 802.15.4 frames, each fragment carries a part of the original IPv6 packets. The IEEE 802.15.4 frame has a maximum packet size of 128 bytes; instead, the IPv6 header size is 40 bytes, UDP and ICMP header sizes are both 4 bytes, fragmentation header adds another 5 bytes overhead. Without compression, 802.15.4 is not possible to transmit any payload effectively.

Routing: There are many existing routing protocols in 6LoWPAN like 6LoWPAN Ad-hoc On-Demand Distance Vector (LOAD), Multipath-based 6LoWPAN Ad-hoc On-Demand Distance Vector (MLOAD), Dynamic MANET On-Demand for 6LoWPAN Routing (DYMO-Low), Hierarchical Routing (Hi-Low), Extended Hi-Low and Sink Ad hoc On-Demand Distance Vector Routing (SAODV), etc. Note that such routing action is

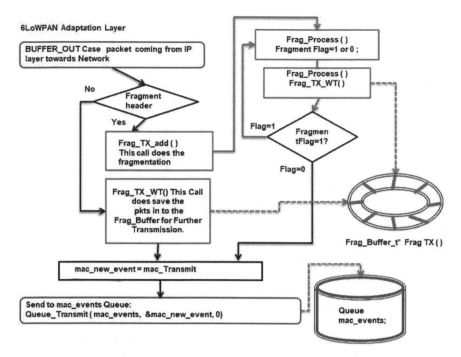

Figure 14.6 6LoWPAN adaptation layer BUFFER_OUT case message flow.

taken care of at the IP layer and hence the fragmentation layer shall precede the routing layer.

In the above-shown flow Figure 14.6, a packet is coming from the application layer toward the network for transmission. If there is no requirement of the fragmentation, then the data packet pointer is added to the **mac_events** queue and indication is sent to the MAC layer for further processing. If fragmentation has to be done, one may have to do a maximum of 16 fragments based on about 80 bytes payload and 1280 bytes minimum packet size. In simple terms, the call **Frag_Process**() takes care of this. Each fragmented packet is kept in the TX ring buffer and the pointers shall be sent to the **mac_events** queue for further processing at the lower layer.

In the above shown flow (Figure 14.7), a packet is coming from network toward application layer as part of the reception. If there is no requirement of fragmentation, then the data packet pointer is added to the **IP_events** queue and indication is sent to the IP layer for further processing. If de-fragmentation has to be done, the call **De_Frag_Process** () takes care of this. Each de-fragmented packet is kept in the RX ring buffer and the pointers shall be sent to the **IP_events queue** for further processing at the upper layer.

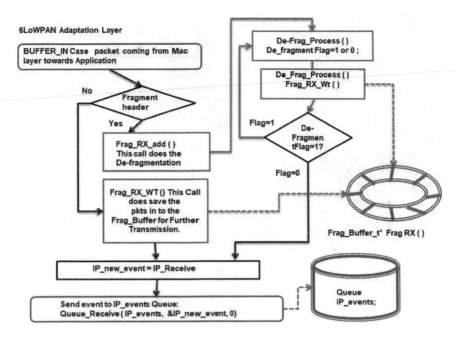

6LoWPAN Adaptation Layer

BUFFER_IN Case packet coming from Mac layer towards Application

Fragment header

No

Yes

Frag_RX_add ( ) This call does the De-fragmentation

Frag_RX_WT () This Call does save the pkts in to the Frag_Buffer for Further Transmission.

De-Frag_Process ( ) De_fragment Flag=1 or 0 ;

De_Frag_Process ( ) Frag_RX_Wt ( )

De-Fragmen tFlag=1?

Flag=1

Flag=0

Frag_Buffer_t* Frag RX ( )

IP_new_event = IP_Receive

Send event to IP_events Queue: Queue_Receive (IP_events, &IP_new_event, 0)

Queue IP_events;

Figure 14.7 6LoWPAN adaptation layer BUFFER_IN case message flow.

14.5 6LoWPAN Network (IP) Layer

At the network layer, 6LoWPAN packets can be created in two ways: full-function IPv6 using routing-over routing mode wherein fragmentation of packets is possible and mesh-under routing in which case (under Nanostack and FreeRTOS) no fragmentation of packets is possible. In the case of Contiki OS, one can implement both cases if need be. But in the case of FreeRTOS and Nanostack, designers can implement only mesh-under routing very quickly and easily. Packets can reach the IP layer either from the MAC layer (**BUFFER_IN case, i.e., toward the application layer**, [refer Figure 14.8]) or from Transport layer (**BUFFER_OUT case, i.e., toward the network**, [refer Figure 14.9]).

To facilitate clean operation, one can create two numbers of Ring buffers: one for incoming packets (BUFFER_IN case) and the other for the outgoing packets (BUFFER_OUT case).

The 6LoWPAN network layer provides the internetworking capability to sensor nodes. The main considerations of this layer are addressing mapping and routing protocols. As described earlier, mesh-under routing decision occurs in the 6LoWPAN adaptation layer. On the other hand, route-over routing decision occurs in the 6LoWPAN network layer. The main

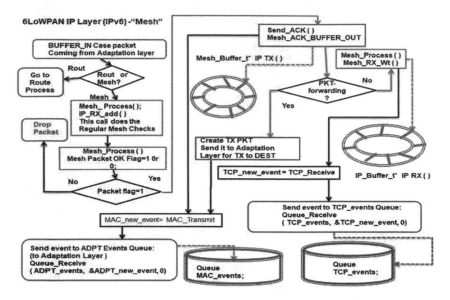

Figure 14.8 6LoWPAN IP layer mesh BUFFER_IN case message flow.

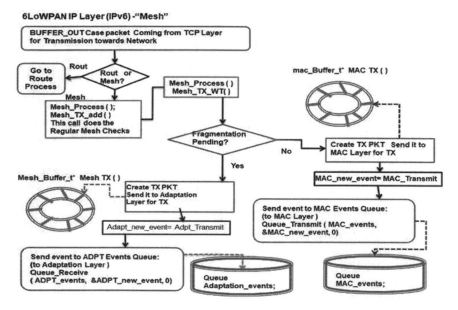

Figure 14.9 6LoWPAN IP layer mesh BUFFER_OUT case message flow.

differences of routing issues in route-over and mesh-under schemes are in the packet/fragment forwarding process rather than the route establishment phase. In the route-over scheme, each link-layer hop is an IP hop and each node acts as an IP router. The packet is forwarded hop by hop from source to destination between these links. The packet's payload is encapsulated in the IP header.

IP packet is to be fragmented and the fragments are to be sent to the next-hop based on routing table information, and if the 6LoWPAN Adaptation layer is present, then the fragmentation process is completed at that layer. If the adaptation layer in the next hop receives all the fragments success-fully, then it creates an IP packet from fragments and sends it up to the network layer. After that, the network layer sends the packet to the upper layer (**transport layer**), if the packet is destined for it. Otherwise, it forwards the packet to the next hop according to the routing table information. However, if some fragments are missing, all fragments are retransmitted to a one-hop distance. In mesh-under routing schemes, a one-hop fixed-hop, or broadcast model is used. While in route-over scheme, complete IP routing algorithms are used and one can use all possible route schemes depending on the designer's choice.

In the case of the mesh routing and BUFFER_IN case, as shown above, (Figure 14.8), the process is similar. Note that in this case it is assumed that the 6LoWPAN Adaptation layer is present also. If the Adaptation layer is not present, then the ACK_BUFFER_OUT shall be directly sent to the MAC layer.

In the case of the BUFFER_OUT case (Figure 14.9) and Mesh routing scenario, we shall do the check for the fragmentation layer, and if it is pre-sent, we shall send the packet for fragmentation. If the fragmentation layer (i.e., 6LoWPAN Adaptation layer) is not present, then the packet is directly sent to the MAC layer.

At Mesh layer **Mesh_Process** () will have the necessary actions as per the Mesh protocol defined in RFC 4944. Routing information is provided in the Mesh layer through two main tables: Neighbor Table and Routing Table. The Neighbor Table contains contact information related to nodes that were heard in the past, either by broadcast or by unicast packets. The Routing Table contains contact information for nodes that are not neighbor nodes. New entries in this table are added with callings to **Update_Routing_Table**() setting a value on **buf. event** (packet reception or packet transmis-sion is known as **buf. event**) for cases other than **REMOVE_ROUTE** and **ROUTE_ERR** in the instance **buf**.

In BUFFER_IN case, packet has come to Mesh layer, and if the destin-ation is neighbor, then it is a packet forwarding case(**Check_Neighbor_Table**(), **call**), and hence accordingly it is again sent to the transmission buffer and attached to the BUFFER_OUT and sent to the Fragmentation layer or MAC layer.

In the BUFFER_OUT case when a packet has come from the UDP layer for transmission, **Mesh_Process** () checks if the destination address of the packet to be sent matches a previously stored neighbor node (Check_ Neighbor_Table(), call). If this is the case, the packet is pushed to the queue events for further processing in the next lower layer in the stack (MAC layer).

If the destination address of the packet being sent (BUFFER_OUT case) does not belong to a previously stored neighbor, the logic checks the routing table by calling **Check_Routing_Table**(). If an entry in the Routing Table exists, the 6LoWPAN Mesh-Header is added to the packet with a source address equal to the address of the node and a mesh final *address* equal to the intended destination address. The destination address in the MAC layer is set to the address found in the routing table.

If an entry in the Routing Table does not exist, the logic verifies whether there is a neighbor with low Received Signal Strength Indication (RSSI)—a state previously retrieved by **Check_Neighbor_Table**(). Despite the existence of a neighbor with low RSSI or not, the logic pushes the packet to the MAC layer. The packet may be sent to the MAC layer to be broadcasted—depending on the logic designer's choice, and in this specific case, it shall reach the 6LoWPAN Border Gateway (**LBR**) also so that the destination can be properly verified and the packet is sent accordingly.

As shown in Figure 14.10 BUFFER_IN case, if a packet has to be acknowledged, then accordingly related specific control packets shall be sent back to the below lower layer (6LoWPAN Adaptation layer or MAC layer depends on implementation). If the received packet is complete in all respects, then it shall be handed over to the higher layer Transmission Control Protocol (TCP).

In the case of BUFFER_OUT and Routing case (Figure 14.11), the process is similar and all the necessary routing algorithms are added and the number of hops, etc., are fixed according to the available data and checked for the fragmentation requirements and the packet is either sent to the 6LoWPAN Adaptation layer or MAC layer depending on the process.

In the above four diagrams, the conceptual routing process for the IP layer has been shown and it is to be understood that the **IP_Process**() or **Mesh_Process**() will take care of the necessary protocol aspects. If required and convenient, designers may add required additional subfunction calls, to complete the Routing/Mesh process to meet different choices.

As part of Contiki "uIP" (Network layer), implementation is designed to have only the absolute minimal set of features needed for a full TCP/IP stack. It can only handle a single network interface and contains the IP, ICMP, UDP, and TCP protocols. uIP can be written in the C programming language. Many other TCP/IP implementations for small systems assume that the embedded device always will communicate with a full-scale TCP/IP implementation running on a workstation-class machine. Under this

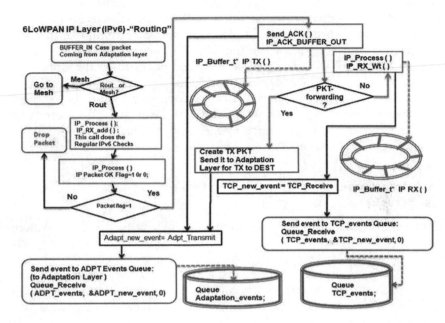

Figure 14.10 6LoWPAN IP layer routing—BUFFER_IN case message flow.

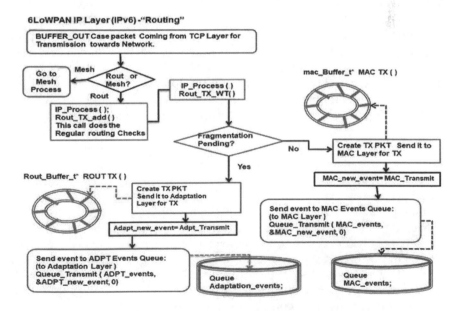

Figure 14.11 6LoWPAN IP layer routing—BUFFER_OUT case message flow.

assumption, it is possible to remove certain TCP/IP mechanisms that are very rarely used in such situations. Many of those mechanisms are essential; however, if the embedded device is to communicate with another equally limited device, for example, when running distributed peer-to-peer services and protocols, uIP is designed to be RFC compliant to let the embedded devices act as peer and standard network citizens. The uIP that is Contiki TCP/IP, implementation is not tailored for any specific application.

All the necessary uIP functions (**Contiki IP layer and Transport layer TCP or UDP**) can be implemented by the way of calling the necessary uIP functions under **IP_Process ()**.

14.6 Implementation of uIP Layer (FreeRTOS and Nanostack)

Application Events: The application must be implemented as a C function, **UIP_APPCALL()**, that uIP calls whenever an event occurs. Each event has a corresponding test function that is used to distinguish between different events. The functions are implemented as C macros that will evaluate to either zero or non-zero. Note that certain events can happen in conjunction with each other (i.e., new data can arrive at the same time as data is acknowledged).

The Connection Pointer: When the application is called by uIP, the global variable **uip_conn** is set to point to the **uip_conn structure** for the connection that is handled currently, and is called the "current connection." The fields in the **uip_conn** structure for the current connection can be used, for example, to distinguish between different services, or to check to which IP address the connection is connected. One typical use would be to inspect the **uip_conn->lport** (the local TCP port number) to decide which service the connection should provide. For instance, an application might decide to act as an HTTP server if the value of **uip_conn->lport** is equal to 80 and act as a TELNET server if the value is 23.

Receiving Data: If the uIP test function **uip_newdata()** is non-zero, the remote host of the connection has sent new data. The **uip_appdata** pointer points to the actual data. The size of the data is obtained through the uIP function **uip_datalen()**. The data is not buffered by uIP but will be over-written after the application function returns, and the application will there-fore have to either act directly on the incoming data or by itself copy the incoming data into a buffer for later processing.

Sending Data: When sending data, uIP adjusts the length of the data sent by the application according to the available buffer space and the current TCP window advertised by the receiver. The amount of buffer space is dictated by the memory configuration. It is therefore possible that all data sent from the application does not arrive at the receiver, and the application may use the **uip_mss()** function to see how much data that actually will be sent by the stack.

The application sends data by using the uIP function **uip_send()**. The **uip_ send()** function takes two arguments: a pointer to the data to be sent and the length of the data. If the application needs RAM space for producing the actual data that should be sent, the packet buffer (**pointed to by the uip_ appdata pointer**) can be used for this purpose.

The application can send only one chunk of data at a time on a connection and it is not possible to call **uip_send()** more than once per application invocation; only the data from the last call will be sent.

Retransmitting: Data retransmissions are driven by the periodic TCP timer. Every time the periodic timer is invoked, the retransmission timer for each connection is decremented. If the timer reaches zero, retransmission should be made. As uIP does not keep track of packet contents after they have been sent by the device driver, uIP requires that the application takes an active part in performing the retransmission. When uIP decides that a segment should be retransmitted, the application function is called with the **uip_rexmit()** flag set, indicating that a retransmission is required.

The application must check the **uip_rexmit()** flag and produce the same data that was previously sent. From the application's standpoint, performing retransmission is not different from how the data originally was sent. Therefore, the application can be written in such a way that the same code is used both for sending data and retransmitting data. Also, it is important to note that even though the actual retransmission operation is carried out by the application, it is the responsibility of the stack to know when the retransmission should be made. Thus, the complexity of the application does not necessarily increase because it takes an active part in doing retransmissions.

Closing Connections: The application closes the current connection by calling the **uip_close()** during an application call. This will cause the connection to be cleanly closed. To indicate a fatal error, the application might want to abort the connection and does so by calling the **uip_abort()** function. If the connection has been closed by the remote end, the test function **uip_closed()** is true. The application may then do any necessary cleanups.

Reporting Errors: two fatal errors can happen to a connection, either that the connection was aborted by the remote host, or that the connection retransmitted the last data too many times and has been aborted. uIP reports this by calling the application function. The application can use the two test functions **uip_aborted()** and **uip_timedout()** to test for those error conditions.

Polling When a Connection Is Idle: uIP polls the application every time the periodic timer fires. The application uses the test function **uip_poll()** to check if it is being polled by uIP. The polling event has two purposes. The first is to let the application periodically know that a connection is idle, which allows the application to close connections that have been idle for too long. The other purpose is to let the application send new data that has been produced. The application can only send data when invoked by uIP, and therefore the poll event is the only way to send data on an otherwise idle connection.

Listening Ports: uIP maintains a list of listening TCP ports. A new port is opened for listening with the **uip_listen**() function. When a connection request arrives on a listening port, uIP creates a new connection and calls the application function. The test function **uip_connected**() is true if the application was invoked because a new connection was created. The application can check the "uIP port" field in the **uip_conn** structure to check to which port the new connection was connected.

Opening Connections: New connections can be opened from within uIP by the function **uip_connect**(). This function allocates a new connection and sets a flag in the connection state which will open a TCP connection to the specified IP address and port the next time the connection is polled by uIP. The **uip_connect**() function returns a pointer to the **uip_conn structure** for the new connection. If there are no free connection slots, the function returns NULL. The function **uip_ipaddr**() may be used to pack an IP address into the two-element 16-bit array used by uIP to represent IP addresses.

Two examples of usage are shown below. The first example shows how to open a connection to TCP port 8080 of the remote end of the current connection. If there are not enough TCP connection slots to allow a new connection to be opened, the **uip_connect**() function returns NULL and the current connection is aborted by **uip_abort**().

```
void connect_example1_app(void) {
if(uip_connect(uip_conn->ripaddr, HTONS(8080)) == NULL) {
uip_abort();
}
}
```

The second example shows how to open a new connection to a specific IP address. No error checks are made in this example.

```
void connect_example2(void) {
uip_addr_t ipaddr;
uip_ipaddr(ipaddr, 192,168,0,1);
uip_connect(ipaddr, HTONS(8080));
}
```

The implementation of this application is shown below. The application is initialized with the function called **example1_init**() and the uIP callback function is called **example1_app**(). For this application, the configuration variable **UIP_APPCALL** should be defined to be **example1_app**().

```
void example1_init(void) {
uip_listen(HTONS(1234));
}
void example1_app(void) {
if(uip_newdata() || uip_rexmit()) {
uip_send("ok\n", 3);
}
}
```

The initialization function calls the uIP function **uip_listen()** to register a listening port. The actual application function **example1_app()** uses the test functions **uip_newdata()** and **uip_rexmit()** to determine why it was called. If the application was called because the remote end has sent it data, it responds with an "ok." If the application function was called because data was lost in the network and has to be retransmitted, it also sends an "ok." Note that this example shows a complete uIP application. It is not required for an application to deal with all types of events such as **uip_connected()** or **uip_timedout()**.

Similarly, all the necessary IP functions (**Nanostack IP layer and Transport layer UDP, under the FreeRTOS**) can be implemented by the way of calling the necessary IP functions under **IP_Process ()**.

14.7 Implementation of fIP Layer (FreeRTOS and Nanostack)

The processing of the packet in the BUFFER_IN direction (a packet is coming from network to Application) comes from the MAC layer to Mesh layer (IP layer). One has to check for the packet headers LOWPAN_HC1, LOWPAN_BC0, and IPv6, as defined in RFC 4944. We need to check whether they are proper and continuous, and then add them into the UDP queue. Also note that there shall be an acknowledgment of the packets through **Time_To_Live** actions, for UDP datagram packets. If frame checks are not proper, one may have to drop the packets. In the BUFFER_OUT direction as the packet is coming from the application toward the network, one has to build the required headers LOWPAN_HC1, LOWPAN_BC0, and IPv6, as defined in RFC 4944, and send the packet for transmission.

At first, we need to verify the **Time_To_Live Flag (TTL_flag=1 or 0)** that is to check whether the packet has reached this node within the set time, in the case of the BUFFER_IN case. Otherwise, if the packet is being transmitted from this node BUFFER_OUT case, we need to set the **Time_To_Live** Flag to one to start the counter.

The second verification is related to the **Link_Lost** that happens when the packet reception is not in time or we lost the consecutive packets that are supposed to be received. In the **Link_Lost case,** the MAC layer adds a flag in the packet that is identified in this conditional branch and triggers an exception message. The exception message is delivered to the application layer for packets addressed to the current node through a dedicated queue to inform exceptions (**queue Event_Queue**), or otherwise delivered using an ICMP message to the destination address set in the packet.

It is to be noted that most of the actions are completed in **Mesh_Process** (), call in the case of the Nanostack, and hence additionally the required exceptions are verified and packet headers are added (BUFFER_OUT case) or headers are verified (BUFFER_IN case).

14.8 6LoWPAN Transport Layer

6LoWPAN transport layer is responsible for process-to-process delivery. It delivers data segments to the appropriate application process on the host computers. In this layer, there are two types of transport protocols: UDP and TCP. On the source side, either TCP or UDP connections are established based on the application. For example, with the FreeRTOS and Nanostack, it is mostly a UDP process along with an optional Network Routing Protocol (NRP) process. Even though NRP comes over and above the UDP, for all practical purposes both layers may be implemented together. While with Contiki OS it is either TCP process or UDP process depending on the designer's choice and some designers may implement CCN instead of full Border Gateway Protocol (BGP) or both. The data from the application layer is pushed to (after creating proper sockets at the application layer) either UDP or TCP segments and attached to create the required protocol process (TCP or UDP processes). At the destination side, after the UDP or TCP segments are received from the network layer, the transport layer processes the segment based on the protocol used and sends it up to the application layer. However, the most common protocol applied with 6LoWPAN is the UDP. In the aspect of performance, efficiency, and complexity, TCP is not preferably used with 6LoWPAN.

14.9 Network Routing Protocol (NRP)

NRP is generally used to make IoT device LC that is Full Function Device (FFD) or Gateway for a set of IoT nodes or hosts. NRP works like a gateway between WSN and the remaining world beyond this. LC generally forwards the packets to the LBR (6LoWPAN Border Router) and LBR which is capable of using BGP, etc., required gateway protocols shall send the packets to the central servers for storing the information and further analysis.

NRP uses serial communication (**standard UART protocol**) to receive the packets from the Application layer and hence it is very useful to implement NRP on a PC or workstation to act like the LC. The NRP task **NRP_Task()** handles the following two cases—incoming packets (**from Network to Application**) and outgoing packets (**from Application to Network**). Outgoing packets shall be using a serial interface to the UDP layer, and then through the stack to the Radio for transmission onto the network. Packets that are incoming from the Network shall be loaded into the UDP level via stack and use the same serial interface to reach the application layer. In either case, Transport layer (UDP process) places the packet into the required TX or RX ring buffer, for further actions.

14.10 Contiki Stack of Transport Layer

As we all know, the designer may need to implement either TCP process or UDP process or even both processes that depend on the designer's target IoT device plan.

14.10.1 TCP Functions Which May Be Internal to TCP Process TCP_Process ()

void tcp_attach (struct uip_conn * *conn*, void * *appstate*);

> tcp_attach function attaches the current process to a TCP connection. Each TCP connection must be attached to a process for the process to be able to receive and send data. Additionally, this function can add a pointer with a connection state to the connection.

> Parameters: **conn**—A pointer to the TCP connection; and **appstate**—An opaque pointer that will be passed to the process whenever an event occurs on the connection.

void tcp_listen (u16_t port);

> tcp_listen function opens a TCP port for listening. When a TCP connection request occurs for the port, the process will be sent a tcpip_ event with the new connection request.

> Parameters: **port**—The port number in network byte order.

void tcp_unlisten (u16_t port);

> tcp_unlisten function closes a listening TCP port. Parameters: **port**— The port number in network byte order.

> Note: Port numbers must always be given in network byte order. The functions HTONS() and htons() can be used to convert port numbers from host byte order to network byte order.

struct uip_conn* tcp_connect (uip_ipaddr_t *ripaddr, u16_t port, void * appstate);

This function **tcp_connect** opens a TCP connection to the specified port at the host specified with an IP address. Additionally, an opaque pointer can be attached to the connection. This pointer will be sent together with uIP events to the process.

Note: The port number must be provided in network byte order so a conversion with HTONS()is usually necessary.

This function will only create the connection. The connection is not opened directly. uIP will try to open the connection the next time the uIP stack is scheduled by Contiki.

Parameters:
ripaddr—Pointer to the IP address of the remote host.
port—Port number in network byte order.
appstate—Pointer to application-defined data.

Returns:
A pointer to the newly created connection, or NULL if memory could not be allocated for the connection.

void tcpip_poll_tcp (struct uip_conn * conn);

This function **tcpip_poll_tcp** causes uIP to poll the specified TCP connection. The function is used when the applicationhas data that is to be sent immediately and does not wish to wait for the periodic uIP polling mechanism.

Parameters:
conn—A pointer to the TCP connection that should be polled.

void tcpip_poll_udp (struct uip_udp_conn * *conn*);

tcpip_poll_udp function causes uIP to poll the specified UDP connection. The function is used when the application has data that is to be sent immediately and does not wish to wait for the periodic uIP polling mechanism.

Parameters:
conn—A pointer to the UDP connection that should be polled.

14.10.2 UDP Functions Which May Be Internal to UDP Process UDP_Process ()

#define udp_bind(conn, port) uip_udp_bind(conn, port)

udp_bind function binds a UDP connection to a specified local port. When a connection is created with **udp_new()**, it gets a local port

number assigned automatically. If the application needs to bind the connection to a specified local port, this function should be used.
Note: The port number must be provided in network byte order so a conversion with HTONS()is usually necessary.

Parameters:
conn—A pointer to the UDP connection that is to be bound.
port—The port number in the network byte order to which to bind the connection.

void udp_attach (struct uip_udp_conn * conn, void * appstate)

udp_attach function attaches the current process to a UDP connection. Each UDP connection must have a process attached to it, for the process to be able to receive and send data over the connection. Additionally, this function can add a pointer with a connection state, to the connection.

Parameters:
conn—A pointer to the UDP connection.
appstate—An opaque pointer that will be passed to the process whenever an event occurs on the connection.

struct uip_udp_conn * udp_new (const uip_ipaddr_t * ripaddr, u16_t port, void * appstate)

uip_udp_conn * **udp_new** function creates a new UDP connection with the specified remote endpoint.
Note: The port number must be provided in network byte order so a conversion with HTONS()is usually necessary.

Parameters:
ripaddr—Pointer to the IP address of the remote host.
port—Port number in network byte order.
appstate—Pointer to application-defined data.
Returns: A pointer to the newly created connection, or NULL if memory could not be allocated for the connection.

uip_udp_conn * udp_broadcast_new (u16_t port, void * appstate)

uip_udp_conn * **udp_broadcast_new** function creates a new (link-local) broadcast UDP connection to a specified port.

Parameters:
port—Port number in network byte order.
appstate—Pointer to application-defined data.

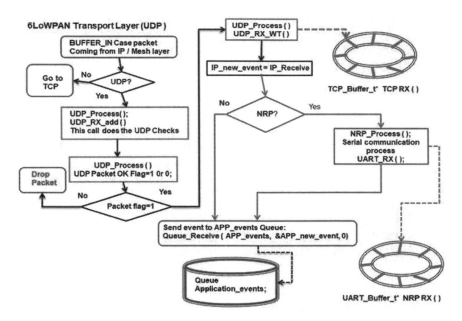

Figure 14.12 6LoWPAN transport layer, UDP BUFFER_IN case message flow.

Returns: A pointer to the newly created connection, or NULL if memory could not be allocated for the connection.

void tcpip_poll_udp (struct uip_udp_conn * conn)

tcpip_poll_udp function causes uIP to poll the specified UDP connection. The function is used when the application has data that is to be sent immediately and does not wish to wait for the periodic uIP polling mechanism.

Parameters:
conn—A pointer to the UDP connection that should be polled.

NOTE: In this document we are making the designers learn the flow of actions at every step and the necessary conceptual understanding of the data packet flow. We are not completely detailing the Contiki OS or FreeRTOS and the related Nanostack or Contiki stack. We hope that designers shall learn about the specific operating system along with stack thoroughly before attempting any IoT design. Please be assured that one cannot miss the required design steps if this document is followed.

In the above we have shown (Figure 14.12) the chronological packet flow involved for the **BUFFER_IN** case at **UDP_Process**(); the UDP process does the standard packet validation, and then checks for if the NRP is involved. It is a case for the FreeRTOS and Nanostack.

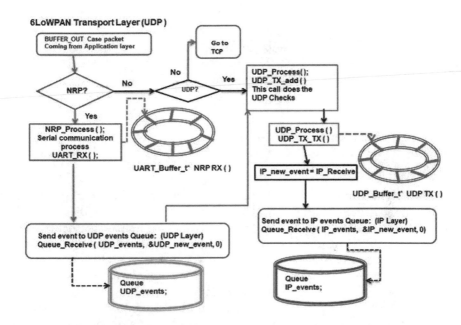

Figure 14.13 6LoWPAN transport layer, UDP BUFFER_OUT case message flow.

Note that those designers who are using Contiki need not check for NRP, but they may need to check for the CCN model within the TCP or UDP processes.

In the above we have shown (Figure 14.13) the chronological packet flow involved for the **BUFFER_OUT** case at **UDP_Process();** UDP checks for if the NRP is involved for the FreeRTOS and Nanostack. If NRP is true, then the necessary UART serial communication process is invoked and then once the bit-by-bit packet is received onto the **UART_Buffer**, the **UDP_events** queue is invoked, for the UDP process-related actions, and then only the packet is sent to the IP layer. If the designer is using Contiki, he need not check for NRP, but they may need to check for the CCN (model within the TCP or UDP processes.

NOTE: In the above, we have shown the UDP process only. But the important functions related to the TCP process are also discussed. We hope that by this time designers would have become familiar with flowcharting each layer of the stack with required functional calls. It is assumed that each designer shall become well acquainted with the TCP process and hence it is not shown here.

14.11 6LoWPAN Application Layer

6LoWPAN application layer uses a socket interface for a specific application. Each 6LoWPAN application opens a socket which is then used to receive or send packets. Each socket is associated with a protocol, TCP or UDP, and source and/or destination ports.

Socket Requirements					
	Protocol	Local-Addr	Local-Port	Foreign-Addr	Foreign-port
Connection-Oriented Server	Socket ( )		Bind ( )	Port ( )	
Connection-Oriented Client	Socket ( )		Connect ( )		
Connectionless Server	Socket ( )		Bind ( )	Recvfrom ( )	
Connectionless Client	Socket ( )		Bind ( )	Sendto ( )	

Figure 14.14 Application layer socket requirements.

Network connections and Socket interfaces:

The application interface is the interface available to the programmer for using the communication protocols. Earlier sockets used to be the means of using IP to communicate between machines. Let us discuss the socket API. With sockets, the network connection can be used as a file. To communicate between two processes, the two processes must provide the information used by ICP/IP (or UDP/IP) to exchange data. This information is the five-tuple: {protocol, local-addr, local-process, foreign-addr and foreign-process} (Figure 14.14). Several network systems calls are used to specify this information and use the socket. The Internet Protocol breaks all communications into packets, finite-sized chunks of data that are separately and individually routed from source to destination. IP allows routers, bridges, etc., to drop packets, but may be retransmitted based on protocol.

Network I/O sockets are asymmetric, the connection requires the program to know which process it is, the client or the server. A connection-oriented network connection is somewhat like opening a file. A connectionless protocol doesn't have anything like an open or close UDP. A network application needs additional information to maintain protections, for example, of the other process. There are more parameters required to specify network connection than the File Input/Output. The parameters have different formats for different protocols. The network interface must support different protocols. These protocols may use different size variables for addresses and other fields. Below are the structures of the different sockets—Generic sockets, IPv4 sockets, and IPv6 sockets. One may be using generic sockets and IPv6 sockets in our case.

/*Generic Socket Address Structure, length=16 */

```
<sys/socket.h>
Structsocketaddr
{
    unit8_t sa_len;
    sa_family_tsa_family; /*address family: AF_XXX value */
    char sa_data[14] /* up to 14 types of Protocol-specific addresses */
};
```

/*IPv4 Socket Address Structure, length=16 */

```
<netinet/in.h>
structin_addr
{
    In_addr_ts_addr; /*32-bit IPv4 address, network byte ordered */
};
structsockaddr_in
{
    unit8_t sin_len; /*length of structure (16 byte) */
    sa_family_tsin_family; /* AF_INET */
    in_port_tsin_port; /* 16-bit TCP or UDP port number, network
    byte ordered */
    structin_addrsin_addr; /* 32-bit IPv4 address, network byte
    ordered */
    charsin_zero[8]; /* unused – initialized to all zeros */
};
```

/* IPv6 Socket Address Structure, length=24 */

```
<netinet/in.h>
Struct in6_addr
{
    Unit8_t s6_addr[16]; /* 128-bit IPv6 address, network bye
    ordered */
};
#define SIN6_LEN /* required for compile-time tests */
Struct sockaddr_in6
{
    unit8_t sin6_len; /* length of the structure (24 byte) */
    sa_family_t sin6_family; /* AF_INET6 */
    in_port_t sin6_port; /* 16-bit TCP or UDP port number, network
    byte ordered */
};
```

It is to be noted that through the socket interface, any two applications can be interfaced through the communication network, that is, two different applications can exchange data. In TCP socket calls, the client (application 1) sends the request to the server (application 2) and the server performs all the functions, that is, socket(), bind(), listen(), and accept(). In concurrent servers, multiple clients (i.e., multiple nodes, routers, bridges, etc.) can be handled at the same time, whereas in an iterative server, the server processes only one request before accepting the next one. In our case, LC—Local Controller and LBR—6LoWPAN Border Router may handle more than one concurrent connection depending on the designer's choice, while node or host can handle one request at a time, for ease of implementation. Some of these below function call examples are taken from UNIX Network Programming by W. Richard Stevens, Prentice-Hall, Englewood Cliffs, NJ, 1997, for ease of understanding and continuity.

14.11.1 Socket Function Calling

```
#include <sys/types.h>
#include <sys/socket.h>
/* Socket Function */
int socket (int family, int type, int protocol);
```
Family: specifies the protocol family {AF_INET for TCP/IP}
Type: indicates communications semantics
SOCK_STREAM stream socket TCP
SOCK_DGRAM datagram socket UDP
SOCK_RAW raw socket
Protocol: Set to 0 except for raw sockets
Returns on success: socket descriptor {a small nonnegative integer}
on error: -1

Example:
```
if ((sd= socket (AF_INET, SOCK_STREAM, 0)) < 0)
err_sys("socket call error");
```

Connect Function
```
int connect (int sockfd, conststructsockaddr*servaddr, socklen_taddrlen);
```

sockfd: a socket descriptor returned by the socket function.
*servaddr: a pointer to a socket address structure
addrlen: the size of the socket address structure
The socket address structure must contain the *IP address* and the *port number* for the connection wanted.
In TCP connect initiates a three-way handshake. Connect returns only when the connection is established or when an error occurs.
Returns on success: 0
on error: -1

Example:
```
if (connect (sd, (structsockaddr*) &servaddr, sizeof(servaddr)) != 0)
err_sys("connectcall error");
```

14.11.2 TCP Socket Calls

Bind Function
int bind (int sockfd, conststructsockaddr* myaddr, socklen_taddrlen);

Bind assigns a local protocol address to a socket.
Protocol Address: a 32-bit IPv4 address and a 16-bit TCP or UDP port number.
sockfd: a socket descriptor returned by the socket function.
*myaddr: a pointer to a protocol-specific address.
addrlen: the size of the socket address structure.
Servers bind their "well-known port" when they start.
Returns on success: 0
on error: -1

Example:
If (bind (sd, (structsockaddr *) &servaddr,sizeof (servaddr)) != 0)
errsys("bind call error");

Listen Function
Int listen (int sockfd, int backlog);

Listen is called only by a TCP server and performs two actions:
1. Converts an unconnected socket (sockfd) into a passive socket.
2. Specifies the maximum number of connections (backlog) that the kernel should queue for this socket.
Listen is normally called before the accept function.
Returns on success: 0
on error: -1

Example:
if (listen (sd, 2) != 0)
errsys("listen call error");

Accept Function
int accept (int sockfd, structsockaddr*cliaddr, socklen_t*addrlen);

Accept is called by the TCP server to return the next completed connection from the front of the completed connection queue.
sockfd: This is the same socket descriptor as in the listening call.
*cliaddr: used to return the protocol address of the connected peer process (i.e., the client process).
*addrlen: {this is a value-result argument}
Before the accept call: We set the integer value pointed to by *addrlen to the size of the socket address structure pointed to by *cliaddr;

On return from the accept call: This integer value contains the actual number of bytes stored in the socket address structure.

Returns on success: a new socket descriptor, on error: -1

For acceptance, the first argument sockfd is the listening socket and the returned value is the connected socket.

The server will have one connected socket for each client connection accepted.

When the server is finished with a client, the connected socket must be closed.

Example:
sfd= accept (sd, NULL, NULL);
if (sfd== -1) err_sys ("accept error");

Close Function
int close (int sockfd);

Close marks the socket as closed and returns to the process immediately.

sockfd: This socket descriptor is no longer useable.

Note: TCP will try to send any data already queued to the other end before the normal connection termination sequence.

Returns on success: 0
on error: -1

Example:
close (sd);

In the Contiki OS case, the following socket calls are used:

- #define PSOCK_INIT(psock, buffer, buffersize) /*Initialize a protosocket. */
- #define PSOCK_BEGIN(psock) /* Start the protosocket protothread in a function. */
- #define PSOCK_SEND(psock, data, datalen) /* Send data. */
- #define PSOCK_SEND_STR(psock, str) /* Send a null-terminated string. */
- #define PSOCK_GENERATOR_SEND(psock, generator, arg) /* Generate data with a function and send it. */
- #define PSOCK_CLOSE(psock) /* Close a protosocket. */
- #define PSOCK_READBUF(psock) /* Read data until the buffer is full. */
- #define PSOCK_READTO(psock, c) /* Read data up to a specified character.*/
- #define PSOCK_DATALEN(psock) /* The length of the data that was previously read. */

- #define   PSOCK_EXIT(psock)   /*   Exit   the   protosocket's protothread. */
- #define PSOCK_CLOSE_EXIT(psock) /* Close a protosocket and exit the protosocket's protothread. */
- #define PSOCK_END(psock) /* Declare the end of a protosocket's protothread. */
- #define PSOCK_NEWDATA(psock) /* Check if new data has arrived on a protosocket.
- #define PSOCK_WAIT_UNTIL(psock, condition) /* Wait until a condition is true. */

One can take FreeRTOS-related socket inputs from the "FreeRTOS TCP Tutorial" pages. However, below are some indications.

Socket is created using the **FreeRTOS_socket()** API function with the xType (second) parameter set to FREERTOS_SOCK_STREAM, configured using the **FreeRTOS_setsockopt()** function, and bound to a port using the **FreeRTOS_bind()** function.

If the socket is used to implement a server, then call **FreeRTOS_listen()** to place the socket into the listening state, and call **FreeRTOS_accept()** to accept incoming connections. Source code examples are provided on FreeRTOS TCP tutorial pages.

To change the size of the receive and send buffers used by the TCP socket from their defaults, call **FreeRTOS_setsockopt()** using the FREERTOS_SO_RCVBUF and FREERTOS_SO_SNDBUF parameters, respectively. This must be done immediately after the socket is created and before it is connected.

If **ipconfigUSE_TCP_WIN** is set to 1 in **FreeRTOSIPConfig.h,** then the socket will use a sliding window to minimize overhead and maximize throughput. The size of the sliding window can be changed from its default using the FREERTOS_SO_WIN_PROPERTIES parameter to **FreeRTOS_setsockopt()**. The sliding window size is specified in units of MSS (so if the MSS is set to 200 bytes, then a sliding window size of 2 is equal to 400 bytes) and must always be smaller than or equal to the size of the internal buffers in both directions.

By default, a child socket is automatically created to handle any connections accepted on a listening TCP/IP socket (the default behavior can be changed using the FREERTOS_SO_REUSE_LISTEN_SOCKET parameter in a call to **FreeRTOS_setsockopt()**. Child sockets inherit the buffer sizes and sliding window sizes of their parent sockets.

14.12 6LoWPAN Border Router (LBR)

Now let us understand the big picture of the generic LBR (6LoWPAN Border Router). Note that whenever we say H/W drivers, it is assumed to be included with the required 802.15.4 radio Driver and 80.15.4 PHY in addition to standard few Ethernet and Console interfaces (refer Figure 14.15).

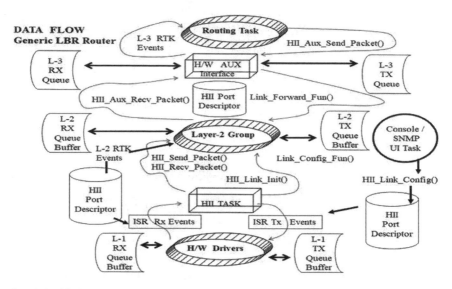

Figure 14.15 Full function generic LBR message flow.

Once the packets are placed in L-1 RX Queue, RX_Event shall be initiated and the data shall reach the next higher level (Hardware Independent Interface (HII) task). When the packet is placed in the L-1 TX Queue, TX_event shall be initiated and the packet is transmitted onto the network.

Once the packet is placed at the L-2 RX Queue, the HII task shall initiate the L-2 RTK events, and then the received packet shall be handled by the L-2 group of protocols (in our case it is the MAC layer). "Layer-2 protocol layer," which is over and above the HII, shall implement the actual protocol part for the remaining existing interface also (Asynchronous PPP, Synchronous PPP, CISCO HDLC, SLIP, LAN LLC, and Frame Relay if any). Above the "Layer-2 protocol layer" is the "Link Support Layer" (LSL). LSL serves as an Interface between Layer-2 group and Layer-3 group layers. Above the Layer-3 group, there shall be the Transport group (TCP, UDP, ICMP, etc.). Then, on top comes the application layer, which may include the User Application Interfaces, Remote Access Servers, etc.

Note: In the case of LBR or equivalent router there is no limitation for memory space and processing power or capabilities compared to LC or Node or Host, and we need to implement more functions that are to be handled concurrently. Hence, one may use Embedded Linux or equivalent OS. To give the complete possible picture, (Figure 14.16) below is the software architectural view of the router in which most of the different components involved with each of the layers are shown for the sake of completion.

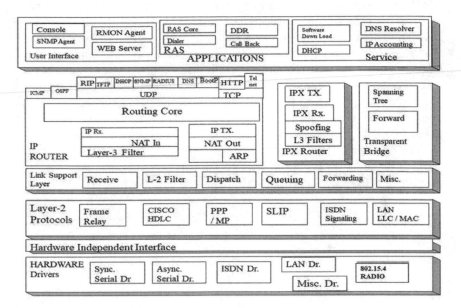

Figure 14.16 Full function LBR or generic router system architecture.

14.13 IoT Design Aspects

Designers should note that the IoT domain will encompass an extremely wide range of technologies, from stateless to stateful events, from extremely constrained to unconstrained aspects, and from hard real-time to soft real-time functions. Therefore, single reference architecture cannot be used as a blueprint for all possible design implementations. While a reference model can probably be identified and maintained by one design team, likely, several reference architectures will co-exist in the IoT.

As an example, the RFID Tag-based identification architecture (using a tree naming mechanism) may be quite different from a sensor-based architecture, which is more comparable to the current Internet. There will also be several types of communications models, such as Thing to Application Server, Thing to Human, or Thing to Thing communication, and clear and complete visualization is a must.

Similar to the Internet evaluation and growth, IoT architecture will evolve with different architectural models, such as the ITU-T model, the NIST model for Smart Grid, the M2M model from ETSI or the Architectural Reference Model from the EU IoT-A project, and related work in other international forums such as the IETF, W3C, etc., should be considered and designers have to always hone their skills along with future developments related to IoT.

IoT devices communicate among themselves and with related services, and the communication is expected to happen anytime, anywhere, and it is frequently done in a wireless, autonomic, and ad hoc manner. Additionally, the services become much more fluid, decentralized, and complex. Consequently, the security barriers in the IoT become much thinner but in a much more disciplined way. It may become much simpler to collect, store, and search personal information and endanger people's privacy. Finally, the concern is rising that control over personal information is increasingly out of the hands of people. This goes beyond the risks people are currently used to, leading to new security requirements. Hence, in due course of time, there shall be some additional legal frameworks and security frameworks to be in-built to the design considerations of IoT, to be remembered.

Chapter 15

IoT Security

I am trying to consolidate some views on IoT security mechanisms, as IoT is growing fast and will touch each one of us in some time to come. Creating a secure model for the IoT requires excellent teamwork, collaboration, coordination, connectivity, and of course understanding each unit of the IoT ecosystem. All the involved devices (or one may call them sensors) and Local Border Router (LBR), etc., must work together and must communicate securely and interact seamlessly with connected systems and infrastructures. For IoT to work securely with rigorous validity checks, authentication, and data verification, the data needs to be encrypted at different levels. As an example, at the application level, software developers need to write code that is stable, reusable, and trustworthy, while following the stipulated international IoT standards and also software code development standards.

It is to be noted that no "Golden unique Key" will fall from heavens to address all IoT security issues, but it is also an agreeable fact that security mechanisms are evolving for more than a quarter-century in this Communications and Computing Industry and establishing their own identity with well-proven methods.

15.1 IoT Value Proposition

Each of the IoT (along with its hardware and in-built software) is always meant for small but unique work, and once deployed IoT does the same work repeatedly for at least two/three years. Depending on its unique deployment requirements and work area, each IoT will send well-planned and known/ expected data, as it does the policing or monitoring job. In general, IoT is designed to send two to ten messages in a minute, which is considered sufficient for most of the work scenarios.

IoT in any systematic deployment scenario will talk to a local host or nearby LBR. Only via LBR, IoT data will reach the respected work-related server(s) or hosts via the Internet. Hence, it is to be noted that IoT device

DOI: 10.1201/9781003303206-15

security mechanisms should be different from that of LBR as LBR is part of a much bigger network.

15.2 IoT Utilization Scenarios

The utilization of a specific IoT is always achieved through its unique deployment requirements and work area—each IoT will send well-planned and known expected data (data in specific limits/or specific pattern) as most of the time IoT does policing or monitoring job. We shall examine the following IoT work areas, just to understand the concept thoroughly:

Car Park: From a specific car park slot, IoT is expected to send whether the specific slot is filled or not with a timestamp. If the specific slot is filled, IoT should send a unique vehicle code with a timestamp, and if that slot is not filled may be "Zeros" at the vehicle code area and timestamp. Note how the unique vehicle code may reach "park slot IoT" via another IoT fixed in the vehicle or maybe by other means like an RFID tag, which we are not discussing here.

Refrigerator or Air Conditioner: From the specific refrigerator or air conditioner IoT is expected to send the unique code of that appliance and whether the optimum temperature is maintained if that appliance is on; if that appliance is off—maybe appliance unique code and some Zeros or Ones will indicate the appliance is off. In general, each air conditioner in India per se has to work in an optimum way between 18° C and 26° C. Hence, if that air conditioner is on, it shall send the temperature data between 18° and 26° because it is powered-on and properly working—which is also the expected data.

With the above two simple examples I am not covering the whole spectrum of the IoT world, but I wish to reiterate that—The utilization of a specific IoT is always achieved through its unique deployment requirements and work area and it is observable fact that each IoT will send well-planned and known expected data. Hence, security may also be achieved from these aspects to be understood by the designer.

IoT Address Mechanism: IoT can be designed to support IPv6, and IPv6 can support 128-bit address patterns, and more than "340 trillion trillion trillion" different addresses, which is very big. If Internet Assigned Numbers Authority (IANA) or any equivalent authority can put in some efforts and allocate a set of addresses only for IoT deployment, it will reduce some confusion in the addressing space; Also such things shall improve the fence of IoT security.

IoT Frame: Let us examine the IEEE 802.15.4 frames (Figures 15.1 to 15.4) for IoT device designer as he is only planning to design a reliable, secure, and properly operating device within the limitations of the standards that are already stipulated by different "Standards Work Groups" (Ex. IETF), to start with an as solid well-established example.

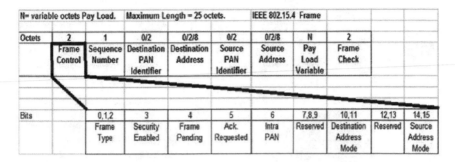

N= variable octets Pay Load.		Maximum Length = 25 octets.			IEEE 802.15.4 Frame				
Octets	2	1	0/2	0/2/8	0/2	0/2/8	N	2	
	Frame Control	Sequence Number	Destination PAN Identifier	Destination Address	Source PAN Identifier	Source Address	Pay Load Variable	Frame Check	
Bits		0,1,2	3	4	5	6	7,8,9	10,11	12,13	14,15
		Frame Type	Security Enabled	Frame Pending	Ack. Requested	Intra PAN	Reserved	Destination Address Mode	Reserved	Source Address Mode

Figure 15.1 IEEE 802.15.4 MTU detail.

Data Frame IEEE 802.15.4, MTU 127 Octets.												
Octets MAC Layer				2	1	0/2	0/2/8	0/2	0/2/8	N	2	
				Frame Control	Sequence Number	Destination PAN Identifier	Destination Address	Source PAN Identifier	Source Address	Pay Load Variable	Frame Check	
Octets	4	1	1	IEEE 802.15.4 MAC Layer - maximum 25 octets header + N octets pay load								
PHY sub Layer	Preamble sequence	Start of frame Delimiter	Frame Length	MPDU								
	SHR		PHR	<============IEEE 802.15.4 MTU 127 Octets ============>								
				PPDU								

Figure 15.2 IEEE 802.15.4 PPDU detail.

Frame format uncompressed IPv6 / UDP, Worst Case Scenario									
			<=========================== MTU of 127 Octets ===========================>						
Preamble sequence	Start of frame Delimiter	Frame Length	MAC Header 23 octets or 44 octets	Dispatch Code	Uncompressed IPv6 Header of total 40 octets	Uncompressed UDP Header of 8 octets	Pay Load of 54 or 33 octets		FCS 2 octets
4 octets	1 octet	1 octet		1 octet					
<==PHY Sub Layer=======>		<====MAC Sub Layer =====>			<==IP Layer (Network) ===> <Transport Layer=>				
1)	Dispatch code (01000001) indicates no compression								
2)	Up to 54 / 33 octets left for payload with a max. size MAC header with null / AES-CCM-128 security								

Figure 15.3 IEEE 802.15.4 UDP packet worst-case detail.

Frame format compressed IPv6 / UDP, Best Case Scenario										
				<================================ MTU of 127 Octets================================>						
Preamble sequence	Start of frame Delimiter	Frame Length	MAC Header 23 octets or 44 octets	Dispatch Code	HC1	IPV6	Uncompressed UDP Header of 8 octets	Pay Load of 92 or 71 octets	FCS 2 octets
4 octets	1 octet	1 octet		1 octet	1 octet	1 octet			
<==PHY Sub Layer=========>		<====MAC Sub Layer =====>				<=IP===>	<Transport Layer=>		

				<================================ MTU of 127 Octets================================>						
Preamble sequence	Start of frame Delimiter	Frame Length	MAC Header 23 octets or 44 octets	Dispatch Code	HC1	HC2	IPV6	UDP 3 Octets	Pay Load of 97 or 76 octets	FCS 2 octets
4 octets	1 octet	1 octet		1 octet	1 octet	1 octet	1 octet			
<==PHY Sub Layer=========>		<====MAC Sub Layer =====>					<=IP===>	<transport>		

1)	Dispatch code (01000010) indicates HC1 compression.
2)	HC1 compression may indicat that HC2 Compression follows.
3)	Maximum compression work for Link-Local addresses and the same does not work for Global addresses.
4)	Any Partially compressed header fields shall be carried after HC1 or HC1/HC2 tags.

Figure 15.4 IEEE 802.15.4 UDP packet best-case detail.

15.3 IoT Security Touch-Points

- First, it should be noted that IoT security is not an add-on part in any way, but it is an integral part of each unit of the IoT ecosystem.
- Optimized IoT device design based on the target deployment scenario shall improve the security. This includes IoT design architecture which should reflect the overall solution. Some of the constraints must be converted as benefits during the design. For example, IoT may not need any human interactions; may be deployed at such locations wherein physical access is difficult; and must use international standards and platforms, to name but a few.
- IoT device hardware and software should adhere to the limitations of stipulated International Standards and respected workgroups. For example, IPv6 addressing mechanisms will improve the device registration and remote provisioning methods with proper controls for the IoT Ecosystem. Proper addressing mechanisms should help to locate device groups, utility regions (human, animal, scientific, simple policing, process control, hazardous environment, device lifetime category, etc.), origin device manufacturer company, etc., to name but a few. Proper addressing mechanisms shall support IoT growth and scaling.
- IoT device boot time security mechanisms should establish the integrity of the embedded software that is running on the device with some digital signature [may be designer company digital signature, or "Power on Self-Test" (POST) mechanisms, etc.] that may

have been generated cryptographically. Such basic trust establishment is the first step of the security fence for each IoT device.

- Building a root of trust and security mechanisms at the System on Chip (SoC) level (hardware and embedded software level, cryptographic algorithms) may create great value for the users (consumers and enterprises) of the IoT devices. Specifically, IoT Devices Power on Self-Test mechanisms should include the onboard Battery Authentication, and check mechanisms which shall enhance the confidence of the users. (Batteries with 10-years + life, Solar, also available)

- Hardware-based security mechanisms outperform software-only-based solutions because of their enhanced protection features. IoT device designers should carefully evaluate and plan such requirements as part of the total IoT ecosystem.

- Selection of IoT OS, IoT middleware, and the protocols shall play a role to deliver a well secure design of IoT. These should help in resolving/fixing constraints like network response time, bandwidth between devices, and meeting regulatory requirements.

- Software updates and security patches for IoT devices may affect millions of devices that are operating in the field. Such updates are to be planned and delivered in such a manner that they consume very little bandwidth and at non-active-working hours depending on the geography and location, and at the same time, care should be taken to ensure that such processes do not compromise the functionality of the targeted IoT device.

- Device provisioning and authentication can improve device security. As part of device authentication, security token methods (symmetric or asymmetric keys) can be planned between the individual devices and the host server. One can use X.509 certificate mechanism also, which shall provide a private key to identify each device. The use of the X.509 certificate method (strong security) or the security token method (less strong security) depends on the security level the designer plans to achieve in the IoT ecosystem. A designer should note that X.509 method allows the authentication of the device at the physical layer as part of the Transport Layer Security (TLS) connection establishment.

- IoT sends regular specific known data patterns as a reply to serve the unique area. IoT is all about data, IoT solution is meant for analyzing the unique location-specific data sending corrections/next actions, and presenting the same to the user group (host server), for analytics. All these steps can be properly planned as part of the IoT solution architecture and hence to improve the security layer too. Securing the data with proper Identity access management shall mitigate the risks.

- It is possible to build IoT frames with AES-CCM-128 security.
- IoT sends its replies/traffic via the respective LBR, which is part of the big network. IoT data mostly at fixed periods, and low rates, hence can be easily analyzed at LBR. Sandbox Technology may be implemented at LBR to enhance the security at LBR and IoT device communication.
- In each IoT frame, the space required for the data payload space is mostly fixed and a part of such left/free space may be used for security mechanisms (encryption).
- IoT data analyzing host server (may sit on Internet or Cloud) which takes decisions based on IoT sent data. It may be safe to assume that the host server shall be located at a big data warehouse (maybe a data center), which includes multiple switches, communication devices, along with multiple cybersecurity arrangements.
- IoT device-generated data is meant for data analytics, and may involve data collection from different sources, different security mechanisms, and different geographies. IoT designers should assure that security mechanisms should ensure the authenticity of data from different sources, and should meet the requirements of targeted application needs, which is the major point for the success of the IoT ecosystem.
- Planning and designing with some security at each device of the IoT ecosystem is the best solution for the overall security enhancement. Some level of security choice and flexibility has to be exercised by the designer after evaluating the overall security requirements of the target IoT requirements and solution. For example, Microsoft views this as IoT Device Security, Connection Security (i.e., security levels at Internet Connections), and Cloud Security because the host sits on the Cloud.
- If IoT devices and the related Border Gateways are built with proper security mechanisms, it shall become difficult for malware designers to attack such multiple fenced areas.
- IoT Host Server: It is a fact that each IoT will send well-planned and known expected data to its work-related server and act based on the instructions received from the host. At server and also at IoT, one can plan a security mechanism around this well-known data pattern, which is expected to be received from IoT.
- Security Mechanism in IPv6 Address Space: This may become possible only if the Internet Address Allocation Authority can fix the addresses for IoT devices as a whole in a systematic way. IoT-device/node hop limit may be set to a well-known value. This is because at each LoWPAN (IEEE 802.15.4) network the number of nodes (IoT devices) can be limited to a known fixed number (32, 64, 128, etc.). It may become mandatory to design the Border

Gateway along with the IoT devices in certain cases to enhance the security mechanisms.

- Part of payload space may be used for security mechanism: In the above IEEE 802.15.4 frames, a serious designer may be able to implement such a plan that periodical data to be transmitted from IoT may fit into fewer bytes or to be specific, less than the payload space available, then the remaining space can be used for the unique security mechanism (soft tags), which can be understood only by the related control Server software.
- Sandbox Technology: Let us discuss traditional Sandbox Technology also, whose concept means that appliances per se LBR can execute and inspect network traffic coming from IoT devices, and run application data, such as Adobe Flash or JavaScript, to uncover malicious code. To stop attacks on sandbox methods, experts developed sandboxing techniques that include "content-aware" features to help ensure malware does not evade sandbox analysis. Sandboxing technology should show awareness of the metadata features of the samples it analyzes—such as determining whether the sample includes any additional action upon closing of the document, etc.

IoT device design architecture should target and may support the following:

- Focused design and integration effort.
- Integrated security mechanisms.
- Protection of intellectual property (IP) and data.
- Protection for the business application models.
- Support quality and safety.
- Should support temperature ranges like −40° C to +85° C.
- Ease of deployment.
- The broad range of use cases, but not at the cost of low security.
- Architecture should enable new feature deployment and protected upgrades.

15.4 IoT Future—Closing Remarks

Since 2020s, manufacturers are creating methods to apply "Things" (IoT) to suit their specific requirements, and such industries are confident of business value. Manufacturers view that combining IoT technology and expertise in specific industrial applications enables faster problem-solving and increased productivity. Shop floor Programmable Logic Controller (PLC) inputs via IoT form the basis for MIS decisions from time to time, and in turn, they will control productivity. It is outfield operations downtime reduction, like oil rigs, supply chine inputs for production, to name but a few.

Major changes are taking place in transport and mobile operations, as one can collect the possible accident situations—in addition to roadblocks and traffic congestion, real-time fleet monitoring, heavy truck driver behavior tracking, (to remove transport delays) re-routing the certain important pre-planned transport (vehicles) onto different routes is a value from IoT deployments.

GE could deploy IoT along with their windmills and could be able to learn and improve their windmill performance, through wind speed forecast, and such will become a regular and necessary feature, in due course, for better power delivery and longevity of the machines deployed outdoor.

Soon, IoT can be a trusted friend to authorize the digital signatures, point of presence recognition (individual identity—with 256-bit security layer or with retina identifier), in casinos, stock markets floors, rig floors, etc. For free of cost, individual entry/exit indicators, will become a major fraud stopper/identifier, and also an attendance recorder. It can also stop working once it loses contact with its kin (the person) to whom that specific IoT is assigned.

The most important IoT application is the point of presence, identification as one goes into a specific area (industry floor, mall floor, rig floor, etc.), that area-related application can be temporarily downloaded into the device, and can be enabled with/without the knowledge of the person, and IoT will create a new method of helping decisions.

Recent smart factory management of Moderna (Biotechnology—Vaccine Manufacturer) and Schneider Electric Facility in the USA, with their IoT deployments, has proven the benefits and there are still many to reap from the IoT ecosystem. With such smart factory initiatives, billions of micro and miniature electronic components will become necessary to create the new ecosystem, which is another major business value. An earlier view of only big can realize a "smart factory" is a notion now, and every small and big manufacturer can get the benefits of IoT technology.

These edge IoT ecosystems will give rise to new computing paradigms, with the availability of large data science tools, will become a major input provider, and fraud stopper, and soon IoT will become an intelligent wearable like a watch for each individual, and multiple devices like a smartwatch, smartphone, etc., will merge into individual personal devices, a second identity to each person. Individuals need not prove their identity every time they visit the ATM/bank/airport/Metro/Tube, etc., as they will get recorded in the city servers of the respective location of any metropolis they visit.

In the future, LiFi will replace the home devices and data transport shall become the new paradigm, which will give rise to a new generation of IoT devices.

Index